the eternal child

the eternal child

An Explosive New Theory of Human Origins and Behaviour

Dr Clive Bromhall

Foreword by Desmond Morris

PRESS

First published in Great Britain in 2003

10 9 8 7 6 5 4 3 2

© Clive Bromhall
Foreword © Desmond Morris

First published by Ebury Press
Random House, 20 Vauxhall Bridge Road, London SW1V 2SA

Random House Australia (Pty) Limited
20 Alfred Street, Milsons Point, Sydney
New South Wales 2061, Australia

Random House New Zealand Limited
18 Poland Road, Glenfield, Auckland 10, New Zealand

Random House South Africa (Pty) Limited
Endulini, 5A Jubilee Road, Parktown 2193, South Africa

Random House UK Limited Reg. No. 954009
www.randomhouse.co.uk
A CIP catalogue record for this book is available from the British Library

ISBN 0091885744

Papers used by Ebury Press are natural, recyclable products
made from wood grown in sustainable forests

Typeset by seagulls
Printed and bound in Great Britain by Clays Ltd, St Ives plc

Picture Credits: Whilst every attempt has been made to clear all permissions, please
contact the publisher with any ommisions or comment. Chapter 1: Naef A (1926) Über
die Urformen der Anthropomorphen und die Stammesgeschichte des Menschenschädels.
Naturwissenschaften 14: 445-452. Chapter 2: The Hulton Archive Picture Collection.
Chapter 7: Clive Bromhall. Chapter 10: Nelson Mandela, Oscar Wilde, Albert Einstein c/o
Corbis; Queen Elizabeth 1st – LMG162282 Portrait of Queen Elizabeth I (1533-1603)
(oil on canvas) by John the Elder Bettes, Leeds Museums and Art Galleries (Temple
Newsam House) UK/Bridgeman Art Library. Chapters 3, 4, 5, 8 and 9: Getty images

This book is dedicated to all those who possess
a childlike delight in observing, without
prejudice, the world around them.

It is most emphatically not dedicated to those who
are intent on forcing others to accept their beliefs.

contents

foreword
by desmond morris

Once in a while a book comes along that changes the way we see ourselves. This is such a book. It is a fresh and exciting re-examination of how the human species evolved to become the dominant animal on this planet. It explains how we can live in huge urban communities, so vast in numbers that no other species could cope with the stress. It examines how our bodies and our behaviour changed over the millennia, as ape became apeman and apeman became modern man. It reveals the massive impact of our increasingly extended childhood – a childhood that, in one sense, now stretches almost to the grave.

Clive Bromhall's thesis is that we became the Peter Pan species – the eternal child. From this starting point he shows us how this has reduced aggression, increased cooperation, led to an emotional state that favours love and marriage as a basic reproductive system, and has given us huge, ever-curious, ever questioning brains.

He does not shirk the exploration of the controversial by-products of this evolutionary process. As offshoots of the general trend, there are many curious developments that have until now defied satisfactory explanation – why, for example, do large numbers of people set up pair-bonds with members of their own sex, with whom they cannot reproduce? Why do biologically odd forms of behaviour, such as sado-masochism, paedophilia or incest appear so often in our modern societies?

Clive Bromhall is not a psychologist, he is a zoologist and, as a result, he brings a refreshingly frank and objective eye to bear on a wide range of human topics that have, in the past, generated so much confusion and outrage. If, after reading this book, you do not have a much clearer picture of what it means to be a human being, I shall be very surprised indeed.

Oxford, September 2002

prologue

At the age of forty, I should confidently be able to call myself a mature male ape. If I was a forty-year-old chimpanzee I would already be more than three-quarters of the way through my life, and if I was a gorilla of the same age I would very soon be dead. In some ways, my appearance is in keeping with a mature status. I have developed hair on my face and around my genitals, my voice is deeper than that of most females, and my arm and leg muscles are considerably better developed than those of adolescents of either sex. Some of my behaviour is also appropriately mature. I have obtained a small territory, attracted a sexual partner and produced offspring – all of which I am willing to protect jealously.

And yet, calling myself 'mature' feels somewhat disingenuous and unjust to other apes. Certain aspects of my behaviour as well as my body are decidedly immature. Like a baby chimpanzee, I regularly seek reassuring embraces from my sexual partner and long to be stroked and caressed by her. During pair-bonding behaviour, I press my mouth against that of my wife, just like a baby primate when it is receiving prechewed food from its mother. Moreover, I have a strong attraction to women's swollen breasts – a physical attribute that, in any other species, merely signals that a female is producing large amounts of nutritional milk. As for my virtually hairless, thin-skinned, flat-faced, big-headed body – this far more closely resembles that of a baby chimpanzee than a mature adult male. All in all, and however much I try to resist, I am drawn inescapably towards the conclusion that I, like all other members of my species, am simply a hugely overgrown baby ape.

This is the starting point for a book that, in all honesty, was not meant to be. Instead of being called *The Eternal Child*, it should perhaps have been called *The Unplanned Child*. As it happens, this alternative title would have been doubly apt, since one of the book's major themes is that many of our species' unique features were merely serendipitous mistakes. Initially, the book was going to be about the evolution of human homosexuality. Having been raised in the era of the Selfish Gene – in which every feature of an animal's biology is seen as benefiting its genes in some way – homosexuality seemed to fly in the face of logic. How could a behaviour possibly exist that hindrered an individual's chances of having babies? And so the hunt began for the origin of this extraordinary behaviour.

It was, however, one of those journeys that was to lead to some extraordinary places. While meandering through scientific territories that were well off my usual beaten track, I kept on stumbling across a recurring theme. But rather than simply pointing towards the origin of homosexuality, this recurring theme seemed to explain a vast swathe of our species' behaviour, and even our curious physical appearance. Homosexuality, it became clear, was merely one of many linked – and, incidentally, defining – features of our species. And so the hunt changed. It became a search to find the origin of the human species.

Albert Einstein proposed that 'the whole of science is nothing more than a refinement of everyday thinking'. Whatever else science may do, it has to make sense, as well as provide an element of understanding to our day-to-day experiences. Hopefully, this is precisely what *The Eternal Child* does, as it sets out to explain how a most bizarre evolutionary process created *Homo sapiens* – the ultimate animal on this planet. On the way, it makes no apologies for challenging virtually every theory that exists about the origins of our species and our most cherished qualities – including love, loyalty and intelligence. From breasts to religion, and big brains to standing upright, there is a single thread that links all human biology together. What is more, it has everything to do with the fact that we are all – but men especially – such big babies.

chapter one

a body of evidence

It could so very easily have happened. All that was needed was for one meteor – one out of countless millions that hurtle around our universe at any one time – to have followed a minutely different trajectory. Rather than smashing into our tiny planet, this meteor could have sailed by without any of the world's inhabitants being aware of its existence. But this was not to be. As it was, a massive lump of extraterrestrial rock, about 170 kilometres across, thumped into Earth with such ferocity that it sent a cloud of dust into the atmosphere so thick that it prevented most of the sunlight from reaching the planet's surface. In the gloomy darkness that resulted, the stench of death soon started to take hold. Starved of light, plants withered and rotted, forming a uliginous slime that festered on the barren soil. The animals too make their own noxious contribution to the putrid miasma. First of all, many of the smaller herbivores, with their need for constant fresh food, died. Then it was the turn of the larger animals. One by one, the larger species succumbed to starvation, their bodies forming large rotting islands in a sea of death. Tyrannosaurus rex, Brontosaurus and Stegosaurus – each the result of millions of years of evolution were all wiped out because a single, albeit massive, rock just happened to collide with our planet.

But what if the meteor had just missed Earth and the dinosaurs had not been annihilated? How would today's animals have survived along-side flesh-devouring predators such as Velociraptor? Perhaps most intriguing of all, how would the great apes – a group that includes our own species as well as orang-utans, chimpanzees and gorillas – have fared? The most probable answer is that humans would have been hunted with particular vigour. This is not simply because our upright ancestors would have been easier to chase than their four-legged relatives – although this would almost certainly have been the case. Rather, it is because compared with the other great apes humans have deliciously soft, delicate and tender bodies that would undoubtedly have been highly prized by the giant bloodthirsty lizards. We are, in short, perfect dinosaur food. Whereas the other apes have a tough leathery skin that is covered in coarse hair, as well as thick bones and a heavily armoured skull – all features that make them far from pleasant to eat – the human body is thin-skinned, relatively fine-boned, with a large juicy brain that is housed in a flimsy skull, and it even has a tasty extra layer of fat just beneath the skin. Humans are to the other great apes what lamb is to mutton, veal is to beef, or a spring chicken is to an old broiler.

The human body is one of life's great mysteries. Although we are unquestionably members of the primate group – along with bushbabies, monkeys and the other great apes – our bodies possess certain features that make us stand so far apart that it seems that we should belong to a major group all of our own. Of the 181 species of primates, we are the only one that stands upright on its two back legs. We are also the only one that has become largely hairless except for the occasional tuft of hair, such as that on the top of our head, under our armpits and around our genitals. We are especially unusual in having a massive brain that is housed in an inordinately large and bulbous skull, as well as in having an extremely flat face, diminutively small teeth and a very short lower jaw. As for our soft tissue, female humans are the only primate to have permanently swollen breasts – regardless of whether they are producing milk – while human males are extremely unusual among primates in lacking a bone in their penis. All in all, the human body is bizarre.

So where did all these curious and aberrant features come from? Did each evolve in response to a different environmental challenge? Or is there a link between them? For decades, scientists have strived to identify a function for each anatomical and behavioural characteristic of every species – however minute this feature might be. It has been seen as something of a failure if a 'purpose' cannot be attributed to each aspect of an animal's body. And yet, using humans as a model, there seem to be distinct problems with this approach. It turns out to be extremely difficult for scientists to come up with a single widely accepted explanation for any of our species' defining physical attributes. The scientific literature is replete with theories on the benefits of, say, upright walking – yet, by some scientists' own admission, these are invariably little more than 'just-so stories'. Despite centuries of speculation there is still no single, unifying theory to explain why humans evolved the way they have.

With this puzzle in mind, it is time to start on the first leg of our journey to find the missing link – the Holy Grail of human emotion – the single element that binds together all our mysterious and highly distinctive characteristics.

FOUR LEGS GOOD, TWO LEGS BAD

Lance Armstrong, multiple winner of cycling's ultimate challenge, the Tour de France, would find life in the saddle much more comfortable if only he had the head of a chimpanzee. I am not referring to the competitive advantages of having chimpanzee's brain, but rather to the way that the human skull attaches to the spine. The problem for Armstrong, as with all people who participate in sports that require their bodies to be horizontal, is the way that the human head is attached to the spine. Human heads uniquely attach to the spine by means of a joint in the base of the skull, just behind the lower jaw. When Armstrong walks upright along the road this is fine, since his skull balances neatly on the top of his spine which allows his face to point forwards. However, when his body is horizontal, as it is when he is pedalling at breakneck speed across the French countryside, it means that his face naturally points downwards and so he has to strain his neck far back to look ahead. For an adult chimpanzee, on the other hand, the racing cycling position is perfect, since its head articulates with its spine bone by means of a joint at the back of the skull. This means that when a chimpanzee's body is horizontal – as it is when standing on all fours – its face points directly forwards. As a matter of fact, all other primates – from spider monkeys to mountain gorillas – would find the cycling position more comfortable than humans, because we are the only species whose skull has undergone this ninety degree rotation that allows us to stand upright and yet still look forward – rather than up into the sky. So why did it happen? Why did our skulls rotate? Why did all the other anatomical changes occur that favour an upright walking posture – such as a rotated hip, and a big toe that is alongside the other toes rather than underneath the foot where it can be used to grip branches?

If asked to describe a scene that summed up everything they knew about our earliest ancestors, the vast majority of people would almost certainly come up with variations on a well-worn image that involves half-stooping, hairy, chimpanzee-like apemen shuffling across the East African savannah, perhaps along a lake edge. Some people might even go so far as to describe how our earliest ancestors first stood upright as a response to the drying up of East Africa's rainforests, and that standing on two legs rather than four was an essential adaptation to surviving on the open plains. Not only could our upright ancestors stride effortlessly across the plains, but they could also use their freed hands to make and use tools. The whole story sounds so very neat and logical. A massive change in the climate – in this case the drying out of Africa – led to a shift from forest to open plains, which in turn led to the emergence of a brand new species of ape complete with its own unique adaptations, including walking upright. But there is one fatal flaw in this scenario. It is that standing on two legs did not appear when our distant ancestors first found themselves out in the open.

In truth, by the time that grassy plains were widespread across East Africa, our distant ancestors had already been standing upright for millions of years. Over the past century, as more bones have been discovered, the date at which our ancestors are thought to have first stood up has gone further and further back. The most recent archaeological evidence, in the form of leg bones of an East African skeleton dubbed Millennium Man because of his discovery in the year 2000, has provided good evidence that our earliest ancestors started walking about on their two back legs as long ago as six million years ago – that is, while they were still living in thick forest or woodland. As for the oldest apeman skulls, at about four million years old, these leave us in little doubt at all that our ancestors were standing up very early in our evolution. Just like Lance Armstrong today, these ancestors' heads were attached to their spines by means of a joint at the base of the skull, rather than at the back – indicating that their heads were perched on top of vertical spines.

Clearly, this creates a major problem for theories that link upright walking with living in an open habitat. It confutes, for example, the theory that suggests that upright walking evolved as a way of reducing the amount of the body that is exposed to the sun's rays on the open plains. Moreover, if this body position was so important to the survival of our ancestors, why is it that savannah baboons manage quite well on all fours despite living in exactly the same habitat? As regards the idea that our ancestors walked upright to allow them to transport tools from place to place, walking upright actually pre-dates the earliest tools by at least two million years.

Now for the biggest problem of all for those who attempt to justify our upright posture in terms of its advantages when living on the ground. Standing upright is a particularly bizarre way of adapting to a terrestrial lifestyle – in fact, no other mammal has ever chosen to walk on two legs, despite a huge number of them living on terra firma. The reason for this rarity is hardly surprising given the facts. Firstly, it saves little if any energy. Experiments that involved strapping an oxygen-measuring device to a chimpanzee when it was walking on all fours and then on two legs, found that there was no difference whatsoever in the energy consumption of these two modes of locomotion. Secondly, it makes life downright dangerous, since two-legged animals are considerably slower at running away from predators. Chimpanzees and baboons can not only run 40 per cent faster than us, they are also far more agile. Even the fastest, anabolic-steroid-pumped Ben Johnsons of the world can manage a mere thirty-seven kilometres per hour, which is roughly half the speed that the larger African carnivores can reach. Against carnivorous predators, which chase at speeds of over sixty kilometres per hour, humans are doomed. In the words of the scientists John Gribbin and Jeremy Cherfas, 'a running man moves at about the same pace as a startled chicken'. Combine this with the fact that our early two-legged ancestors would have been surrounded by close relatives of lions, as well as carnivorous bears and sabre-toothed cats, and walking upright starts to look downright ludicrous.

The problems associated with standing upright do not even end there. It also makes animals more conspicuous to predators, as well as far more vulnerable should they get injured. Losing the use of a leg is considerably more serious for a two-legged than a four-legged animal which can invariably run fast on only three legs. Even with two fully functional legs, it is impossible for a human to keep up with, say, a three-legged cat or dog. Compare this with the risible image of an ill-fated ancestor attempting to hop away from an oncoming lion. To compound the issue, two-legged animals are also at far greater risk of getting injured in the first place because of the enormous strain that they put their feet and lower legs . When humans run, the forces applied to their feet exceed three times the body weight, with the result that they are at far greater risk of twisting or tearing important muscles and tendons than if their weight was spread across four limbs. Add to this the increased likelihood of suffering from lower back problems, hernias, haemorrhoids and other circulatory problems, and upright walking seems to be something to be avoided at all costs. As the revolutionary chant of George Orwell's *Animal Farm* inhabitants declared, 'four legs good, two legs bad'.

For our earliest upright ancestors, the situation would have been particularly bad. Whereas their predecessors could shin up trees with consummate ease, using their four gripping feet, the new two-legged apes, with their flat feet, would have found it considerably more difficult to escape up trees when being stalked or pursued by a predator. With regards feeding on the ground, as anyone who has collected strawberries for an afternoon will appreciate, standing upright makes life extremely uncomfortable as it forces us to spend much of our time getting up, sitting down and bending over. Four-legged primates, on the other hand, can easily move along the ground while simultaneously grabbing plants and feeding at the same time. Since seeds, grasses and roots are thought to have formed the basic diet of our early ancestors – something that is strongly suggested by the structure of their teeth – yet again standing up seems to have made life considerably harder than it would have been if we had remained on all fours.

So what was going on? Standing upright seems to cause nothing but problems and confer no obvious life-transforming advantages at all. Of the plethora of suggestions that have been made to explain our upright posture, not one has managed either to win the support of the majority of scientists or come anywhere near to outweighing the huge disadvantages. It is a conundrum that caused Chris Stringer, one of the world's most eminent anthropologists, to admit bluntly, 'we don't know why it [bipedalism] evolved'.

'CURIOUSER AND CURIOUSER!'

As if it was not bad enough losing much of their speed by standing upright, our ancestors simultaneously lost their most effective weapons. In place of the massive canine teeth of their ancestors, which we know were used for fighting rather than feeding because of their wear patterns, the early upright apes possessed relatively small canines. Their jaws were also much shorter and considerably less powerful than their predecessors. It was an extraordinary development, especially since these apes would have been far more vulnerable to predators than today's chimpanzees which have seriously large canines. Open-savannah baboons have anything but reduced teeth – a large male olive baboon has canines that measure nearly 6cm long that can be used to great effect against predators such as a leopards and lions. Nevertheless our ancestors, who were now faced with ground-dwelling predators like never before, were gradually losing their bodies' only means of defence. Just like the evolution of our upright posture, the general opinion on our diminutive teeth can be summarised by the anthropologists Sherwood Washburn and Russell Ciochon: 'At the present time no theory of the evolution of the canine complex is fully adequate.'

Yet more equally strange features also started to emerge at this time. A fossil recently discovered in Kenya, belonging to a 3.5 million-year-old upright ape, had an almost flat-face. Despite jokes about the likelihood of just about anything having a flat face after being buried under rock for 3.5 million years, it does appear that this ancient upright species had indeed lost most of its chimpanzee-like muzzle. So if, as our teeth suggest, we were eating seeds and a variety of other vegetable matter, why were our faces becoming flat when all other seed-eating primates have heavy jaws and long muzzles? A further change that occurred involved our ancestors' upper arm bones that were becoming more delicate, especially where they joined the shoulder blade.

In conclusion, it seems that sometime around six million years ago, our distant ancestors diverged from apes that could run at great speed on all fours, were equipped with flesh-ripping teeth, and were capable of nimbly evading predators by climbing trees. Over the next couple of million years, it goes without saying that only the very best-adapted apes would have survived the hardships that resulted from the transformation of vast expanses of Africa from luxuriant jungle into dry woodland, scrub and open plains. So what was this magnificent species that evolution created? What outstanding product of natural selection was capable of out-surviving all the other fifty or so species of ape that existed in the jungle before the catastrophic drought? It was a relatively slow, small-toothed, weak-jawed, flat-footed ape that tottered about on two legs, with a flat face resembling that of a chimpanzee that had just collided with a brick wall, and which was significantly disadvantaged when it came to climbing. What an impressive debut for our ancestors. It was certainly brazen, unquestionably risky, but most importantly of all, it went against everything that one would expect of an ape that was moving out of the trees. In the words of Lewis Carroll's Alice as her body began to change shape after she had eaten a magical cake in Wonderland, the evolution of the human body is most definitely becoming 'Curiouser and curiouser!'

THE BABY APE

The story of human evolution is so like that of the classic crime novel. A gun wrapped in a handkerchief … a dead body lying in a pool of blood clutching a suicide note … two glasses of whisky on the sideboard. In the best tradition of mystery stories, all cannot be what it seems. It seems absurd and a third party, someone else, something else, has to be involved. But what is it? Regarding our own species' mysterious beginnings, we have plenty of clues, but not one seems to make sense in its own right. Perhaps there are the occasional moments when upright walking might have helped a bit, such as when carrying fruit or babies from one place to another, or when throwing sticks and stones at predators, but chimpanzees manage to do all these things without suffering the enormous costs of becoming permanently upright. What links all our species' peculiar features together, and binds them into one cohesive process? What is the missing element? To find the answer, rather than gazing optimistically into a crystal ball, we would do far better peering into the womb of a chimpanzee.

The sight is astonishing. Just visible through the murky amniotic fluid inside the womb, or rather the formaldehyde that preserves the unborn chimpanzee in a museum bottle, is a tiny creature that possesses virtually every one of the supposedly 'new' features that the earliest upright apes developed from their four-legged ancestors. The unborn chimpanzee has a head that is attached to its spine by means of an articulation at its base, meaning that its head points forwards when it stands upright – just as in adult humans. It also has pelvic structures that are positioned so that the legs hang vertically down when the spine is upright – just as in adult humans. It even starts off life with its big toes alongside the other toes, rather than at the side of the foot as in adult chimpanzees. In other words, chimpanzees start off life, albeit in the womb, with all the features that are central to humans' upright posture. As for its head, this is unmistakably human in its shape, the tiny unborn chimpanzee having a small jaw, tightly packed small teeth, an amazingly flat face and a big rounded skull – all, once again, just like us. It turns out that our species' anatomy is nothing new at all, but merely that of an unborn ape.

The resemblance between infant chimpanzees and adult humans does not end with the shape of the head and an upright posture. At 18 weeks old, unborn chimpanzees still possess another of our species more curious features – our fleshy lips. Although chimpanzees initially have 'everted' lips – that is, the inner membranous tissues of the lips face outwards – gradually the membranous tissues fold inwards, so that by the time that they are born chimpanzees have lost the fleshy surrounding to their mouth. Humans, on the other hand, retain the highly-immature feature of everted lips for the whole of their lives.

But perhaps the most staggering similarity between the unborn chimpanzees and humans concerns the distribution of hair. At 26 weeks old, an unborn chimpanzee's hair covering is almost identical to that of an adult human. While the vast majority of its body is naked, a chimpanzee at this stage of development nevertheless has a prolific mop of hair on the top of its head, as well as eye-brows, and even a slight beard in the form of long whiskers around its mouth and chin.

Along with upright walking, hairlessness is another human feature that has so far defied convincing explanation. If, as some scientists have suggested, it allows extra heat-loss, then why do we not see naked antelope or cheetahs? Surely, like all other open savannah animals, we would benefit from having a good covering of hair to act as a heat-shield. Tangible proof of the inappropriateness of our lack of hair in a hot climate is provided by people such as the Bedouin of the Arabian desert who resort to piling on extra layers to insulate themselves from the burning sun. Whether an animal lives in the hot or the cold, body hair is invaluable in maintaining a steady body temperature. In his book *Beyond Natural Selection*, Robert Wesson takes a very similar view:

'It is unlikely that the human loss of the coat of hair that protects apes, carnivores, and almost all mammals is adaptive. Hair may provide a hiding place for lice and fleas, but animals with short, dense pelage are not usually overrun with parasites, and the loss of useful hair is the more eccentric because a mop is retained on top where it is not especially needed. Hairlessness, moreover, should be compensated by a reasonably tough hide, not necessarily like that of the rhinoceros but at least comparable to that of the pig.'

Humans, of course, do not have a thick leathery hide. Instead, they have a thin and delicate skin that can be punctured and ripped with remarkable ease. Compare, for example, our ability – or rather lack of it – to run bare-legged through brambles, with that of a dog. Whereas dogs can plunge relatively carefree into undergrowth, we risk excruciatingly painful lacerations if we attempt the same without artificial protection in the form of sturdy trousers. A lack of body hair has other drawbacks too. It also makes it considerably more difficult for babies to hang onto their mothers, a problem that is accentuated in an ape that stands upright and that therefore lacks a flat back on which the baby can sit. Just like so many of our species' other features, including our upright posture and flat face, our delicate hairless skin also seems to provide no logical benefits in terms of adapting us to our environment. On the other hand, our bizarre nakedness makes entire sense as merely one more way that our species has come to resemble an unborn chimpanzee.

As well as the obvious similarities between humans and baby chimpanzees, there are also some fascinating details of our body that reveal our species' infantile origins. Human nipples, for example, remain far lower on the chest than in other adult primates. In all other primates, although the nipples start off in a low position early in life, they gradually migrate upwards as the individual becomes more mature. Our nipples, on the other hand, remain in their prenatal low position. There is also evidence that the human heart and lungs fail to mature completely. The Belgian scientist Jos Verhulst has described a number of ways that our lungs resemble those of infant apes, including an almost complete absence of the 'infracardiac lobe' – a structure that normally develops as ape lungs mature – as well as the retention of lung lobes that are clearly subdivided into smaller lobules. Similarly, our aortic arch – the blood vessel that carries blood from our heart to the rest of our body – retains the simple structure that is typical of immature primates. Verhulst and others have even argued that the human voice box is more reminiscent of that of immature rather than adult chimpanzees.

Every so-called 'unique' aspect of human appearance is merely the result of our species having come to resemble an infant ape. Although we grow to a size that is roughly double that of a chimpanzee, we fail to develop the same mature features that they do. Our development has simply been arrested to stop us from acquiring all the mature features that chimpanzees do. We have, in other words, been infantised. In the scientific literature it is a process known as 'neoteny' – meaning the retention of young features into adulthood.

It is evolution at its most striking and efficient. Far from having to turn one structure into something totally different, such as a leg into a wing, a process that can take millions of years to happen, infantisation merely involves minor alterations to the timing of certain growth events. It can, perhaps, best be envisaged as a process that involves the disabling of those genes that cause immature growth patterns to cease and mature ones to take over. Instead of changing shape substantially as we get older, as 'mature' genetic instructions supersede 'immature' ones, our bodies carry on growing according to the genetic instructions that were, in our distant ancestors, active only at the beginning of their lives. And the result is that, as adults, we now resemble huge versions of our distant ancestors' babies.

In terms of an everyday analogy, the process is like that of a washing machine that gets stuck in an early part of its washing cycle. Although the washing machine carries on working for as much time as normally functioning machines, it remains stuck within the early phase until it finally comes to a halt – albeit only part-way through its cycle. Humans similarly remain stuck at an early phase of the normal developmental cycle. Rather than progressing rapidly through the early growth phases, we linger within these for far longer than normal, until finally we come to a halt and progress no further – again, without us having reached the later stages at all. The overall result is that many structures grow for far longer than they otherwise would, while others fail to make an appearance. Hypothetically, if the infantising brakes were somehow removed, and the growth cycle was allowed to continue unabated, then we should all end up looking like the hapless visitors to Britain in the cult film *An American Werewolf in London* – that is, hairy, four-legged and with a big muzzle.

BABY LONG LEGS

The precise way that the infantising process works is of particular interest in relation to our understanding of yet another great mystery of human evolution – our long legs. As long ago as the 1920s, the Dutch scientist Siegfried Bok published a series of papers in which he first described how structures that emerge late in any animal's development remain in a more infantile state than those that emerge early on. Since then, Jos Verhulst has applied Bok's principle to our own species, to find out what predictions it makes when early growth phases become highly extended. Using Bok's findings, Verhulst proposed that late-appearing structures should not only become proportionately larger than those that appear early in development, but that they should also remain in a relatively more infantised condition. Once again, imagine the washing machine cycle as it gradually gets slower and slower before finally stopping well before its natural conclusion. Although the early phases of the cycle are largely unaffected, those phases that immediately precede the stage at which the cycle stops will last longer and longer as the cycle ineluctably grinds to a halt.

Clearly, Jos Verhulst was keen to explore the possibility that our species' most unusual anatomical features – especially our upright posture and long legs – were entirely a consequence of the infantising process. If so, explanations that require each of these features to have adapted us in some way to our environment become totally redundant. So, which are the structures that appear last in human development? What is more, does the simple exaggeration of these particular structures provide us with human-like, rather than ape-like, body proportions? According to Verhulst, the fit is almost perfect. Our actual body shape is near enough identical to that predicted by Bok's principle.

Humans, like all vertebrates, grow head end first. When we first develop in the murky depths of our mother's womb, our arm buds appear considerably earlier than our leg buds, and when we are born, the front end of our body, including our arms, is far better developed than the back end, including our legs. Naturally, this is particularly useful for animals whose ancestors at least had to cling on to their mother's fur from the moment that they were born. Legs, however, are relatively late-developing structures, and as such their exaggerated size in humans fits perfectly with Verhulst's prediction regarding a prolonged period of growth for late-forming parts of the body in infantised species. But our legs also fit in with Verhulst's and Bok's predictions in remaining significantly more infantile in structure than our arms. This is especially clear in the human foot. Not only does the foot develop for longer than the hand, and thus grow larger, it also retains the immature pattern of having all its digits – the toes – in a straight line. In contrast, the human hand, like both the hands and feet of other primates, initially has all its digits in a straight line, but then as it grows, the first digit – the thumb – gradually moves to a side position so that it can grip against the other digits. Similarly, the sesamoid bones, which are small bones that lie at the base of the first digit, are also less developed in the foot than in the hand.

Within the human foot, the same pattern exists, with late-developing structures becoming exaggerated in size, as well as remaining more infantile. The first digit – or 'big toe' – for example, is not only the last of all the digits to develop, but also that which has become most exaggerated in our species. Moreover, as Jos Verhulst and others have pointed out, the big toe has the most infantile of all the toes' nails, with the toenails becoming progressively flatter and less 'claw-like' from the little toe to the late-developing big toe. In clawed animals, the claw tends to start as a flat nail-like structure and only later becomes curved. The same relationship, between the order of development and the shape of the nail, is seen in the hand, in which the nails become progressively infantile, and flatter, from the early-developing little finger to the late-developing thumb.

It seems that a combination of basic developmental principles, and our extended immature growth period, predicts human anatomy with extraordinary accuracy. Desirable or not, an upright posture involving our unrotated head joint and long back legs were merely an unavoidable consequence of our ancestors becoming infantised.

NOT FOR THE FIRST TIME

In the early 1900s, Professor Walter Garstang was the first zoologist to fully appreciate the significance of the infantising process in evolution. What Garstang realised was that many types of animals naturally have at least two versions of themselves – such as a juvenile form that is ideally suited to one environment, and an adult form that is suited to another. With frogs, for example, the immature version – the tadpole – can be a voracious aquatic predator, while the adult version is adapted to spend much of its time out of the water eating flying insects. Garstang was struck by the likelihood that certain environmental changes could lead to the immature version of an animal being far better adapted than the adult form, in which case evolution might simply favour those individuals that serendipitously failed to mature fully. Far from requiring the evolution of new body parts, the perfectly adapted individual could be created by simply preventing the expression of a few genes that triggered an animal's metamorphosis into the adult form. Of all the evolutionary processes, infantisation is perhaps the most effective for rapid, profound and radical change.

Five hundred million years ago, it was almost certainly the same infantising process that launched the evolution world's greatest animals – those with backbones, the vertebrates – from the small larvae of sea urchin-like animals. The similarity between the tadpole-like larvae of the sea urchins and the most primitive vertebrates, prompted Garstang to suggest that the evolution of the whole vertebrate group – from fish to elephants and humans – started when a mutation arose that allowed a sea urchin 'tadpole' to reproduce without having to develop into the adult form. Sir Alister Hardy, another zoologist who was incidentally Professor Garstang's son-in-law, later summed up the revolutionary importance of this idea:

> Garstang realized, before anyone else, the profound influence which such larval adaptations might have on the course of evolution...Evolution takes the remarkable step – or rather stride – from an apparently lowly sedentary form to become an active free-swimming type: the forerunner of a new race to lead on to the vertebrates – to fish – to man … larval novelties may give rise to a new major line in evolution. A monstrous piece of speculation some might say.

Even among the animals without backbones, there is one spectacularly successful group that also appears to have evolved through the

infantising of their ancestors. It is the insects. As with the vertebrates, there is an astonishing similarity between the most primitive
insects and the larval form of millipedes. The body of an adult
insect consists of a head that is made up of six or seven segments, a
thorax of three segments each of which has a pair of legs, and an
abdomen of about ten segments – in other words, it is almost identical to the larvae of certain millipedes before they develop into
adults with legs all the way down the body. In certain insects, tiny
relic legs are even present in the abdomen – as though in these
species the infantising process has been slightly less effective in halt-
ing the progression to the adult form. Moreover, while adult millipedes have breathing and excretory structures that are very different from those of insects, those of larval millipedes and insects are
extremely similar.

Returning to the evolution of our own species, the effectiveness of
infantising in radically altering the appearance of an animal with minimal alterations to the genes is illustrated by the astonishing genetic
proximity of humans and chimps. Following the pioneering and now
classic experiments of Marie-Claire King and Allan Wilson at the
University of California, Berkeley, it has been found that the DNA
of humans and chimpanzees is between 95 and 99 per cent identical.
It was a discovery that immediately prompted King and Wilson to
suggest the importance of genes that regulate development in the
evolution of the human species:

The intriguing result ... is that all the biochemical methods
agree in showing that the genetic distance between humans and
the chimpanzee is probably too small to account for their
substantial organismal differences ... We suggest that evolution-

ary changes in anatomy and way of life are more often based on changes in the mechanisms controlling the expression of genes than on sequence changes in proteins.

The anatomical features that we see in our first upright ancestors of six million years ago were the result of alterations to genes that control the rate of maturation. They were features that previously existed only in the juveniles of their ancestors, but came to be retained in the adult body. Rather than being unrelated, anatomical novelties such as a head that looked forward when the spine was vertical, small teeth and a flat face were all unavoidable consequences of our ancestors regressing to a permanently infantile condition. Over the course of the next few million years, many more changes were to occur to our ancestors' bodies as a result of the same infantising process. The result is the creature that is staring at this page – you – an ape locked in a body that is destined to remain infantile for the whole of its life. The eternal child.

PARTLY HUMAN

It is time to introduce the most extraordinary, engaging and possibly most enlightening – from the point of view of human evolution at least – of all the world's animals. Discovered as recently as 1929, it is an animal whose name derives from a misspelling, on a shipping crate, of a town in Zaire from which one of the first individuals came. The town was that of Bolobo – and the name given to the bewildered creature that was confined within the crate was a 'bonobo'. It is the second species of chimpanzee, and the reason for mentioning it at this point is that, like us, it retains a number of infantile features throughout adulthood. Like humans, the bonobo is something of a package of infantile characters.

This story of bonobos starts with an immensely popular inhabitant of Amsterdam zoo, called Mafuca. At the time that Mafuca lived in Amsterdam, from 1911 until his death in 1916, bonobos were not recognised as a separate species from chimpanzees and so, since he bore a strong resemblance to all the other juvenile chimpanzees in the zoo, he was put in with the rest of them. But the Dutch scientist Anton Portielje was sure that there something was different about Mafuca, and in his 1916 guide to the zoo he speculated that Mafuca might be a new species. Thirteen years later, following the inspection of what was initially thought to be a chimpanzee skull in a Belgian museum, the German anatomist Ernst Schwarz published a scientific paper in which he declared that he had found a new subspecies of chimpanzee. This was later elevated to the status of an entirely new species – it was the bonobo.

Although they grow nearly as large as chimpanzees, bonobos retain a distinctly infantile appearance for the whole of their lives. They have delicate facial features, rounded skulls, high foreheads, as well as flatter faces, smaller teeth and smaller jaws than chimpanzees, all of which are not only infantile characters but also make them look distinctly more human. In addition, their bodies are slender and considerably more elegant than those of chimpanzees, and they keep their white tailtufts, which chimpanzees lose after weaning age. They also have a magnificent mop of black hair on their heads – just like unborn chimpanzees. Even their voices are infantile, being shrill just like those of infant chimpanzees, and they are incapable of making the loud hooting sounds that are so typical of adult chimpanzees. Comparing the voices of bonobos with chimpanzees, one scientist wrote, 'The difference in timbre between the voices of the two species may well be of the same magnitude as that between a small child and a grown man.' Bonobos simply do not grow up as far, or as fast, as chimpanzees – something that is evident from an early age. For example, at one year old, whereas chimpanzee infants are tottering around and even playing with others, bonobo infants are still clinging tightly to their mother and barely capable of walking at all. It is not until they are a year and a half old that bonobos are sufficiently developed to start playing with others.

But what is perhaps even more interesting about bonobos is their posture, since along with their infantile bodies comes a simultaneous tendency to walk upright. Bonobos walk upright for as much as 8 per cent of the time, which is more than any other great ape. In contrast with chimpanzees, with their hunched back, thick neck, broad shoulders and heavy torso, bonobos are nimble lightweights. Not only is their upper body considerably lighter than that of chimpanzees, but – in accordance with Verhulst's and Bok's predictions regarding the exaggeration of late-developing structures in infantile species – bonobos also have strikingly long legs. What is more, bonobos' heads are also significantly better suited to an upright posture than those of chimpanzees, since their spinal attachment point is considerably further beneath the skull than in chimpanzees. The result is that when bonobos stand or walk upright, their heads naturally sit comfortably on top of their straight spines. Chimpanzees, on the other hand, have to hunch their backs to look directly forwards. In short, seeing a bonobo alongside a chimpanzee is like seeing Rudolf Nureyev alongside Quasimodo.

It seems that, for bonobos as well as humans, being infantised goes hand in hand with having a body that is naturally inclined to an upright posture. Put together, the package of changes that occur when a species is infantised unavoidably favours two-legged walking – even if, as certainly seems to have been the case for our earliest ancestors, it makes life considerably more dangerous than walking on all fours.

It is as though bonobos have headed down a similar route to humans, although not nearly so far. Frans de Waal, professor of primate behaviour at the University of Atlanta and an expert on pygmy chimpanzees, has commented on how 'the upright posture of this aging male [bonobo] ... shows that little evolutionary change would have been required for an ancestor with a bonobo-like anatomy to develop a bipedal [upright] gait'. Indeed, strong evidence that humans have passed through a bonobo-like stage comes from comparisons between the skeletons of bonobos and our ancestors of three to four million years ago. Side by side, bonobos and these ances-tors have amazingly similar bodies in terms of the relative lengths of their body, their arms and their legs. As long ago as 1933, Harold Coolidge predicted the possible significance of the bonobo when he wrote: 'It [the bonobo] may approach more closely to the common ancestor of chimpanzees and man than does any living chimpanzee hitherto discovered and described.' If this is indeed the case, then chimpanzees have reverted back to walking on all fours, while the bonobo – a species that still lives in the dense forest of its ancestors – has retained the ancestral feature of being able to walk comfortably on two legs. It is fascinating to note that one of the suggestions as to why chimpanzees reverted to a four-legged lifestyle is because of their need to adapt to more open, drier habitats, such as savannahs and wood-lands. Presumably this is because of the additional speed and climbing ability, in the event of having to escape from a ground predator, that four-legged walking provides. Clearly this directly contradicts the traditional claim that our ancestors adopted an upright posture to adapt them to exactly the same open habitat. Yet again, this highlights the fact that there is no consensus whatsoever as to what led to our ancestors standing upright.

FACE-TO-FACE WITH OUR PAST

The natives of Tuscany refer to it as '*la posizione angelica*' – the angelic position. The sixteenth-century Tunisian writer Nafzawi placed it at the top of his catalogue of sexual positions in *The Perfumed Garden*. Some traditional Christian groups describe it as the only appropriate or natural position, and its name is derived from Christian missionaries who felt the need to spread its use to societies where other positions predominated. It is the 'missionary position', the most common sexual position of the human species. For centuries, this face-to-face mating position has been used to prove humans' elevated position above the rest of the animal kingdom, as it allows us to gaze lovingly into the eyes of our partner while participating in the glorious act of lovemaking. Biology, on the other hand, has a distinctly less romantic explanation. The simple reason why humans favour the missionary position is because a woman's vulva – the opening of her genital tract – faces towards the front of her body, rather than backwards as in nearly every other mammal. Because of this, human males approach females from the front rather than from behind in order to mate.

According to the normal development of chimpanzee genitals, the sexual organs of adult men and women are staggeringly under-developed for life in the outside world. Just as with so many other human body parts, human male and female sexual organs are far more similar to those of unborn, rather than adult, chimpanzees. In the human female, a major indicator of this retarded development is her forward-facing vulva – something that exists in unborn female chimpanzees but not in adults. As a female chimpanzee develops, her vulva initially faces forwards, but then, as her body grows further, her vulva is gradually pushed backwards between her legs until, by the time she is born, it faces towards the rear when she is standing on all-fours.

It is not only the position of women's genitals that illustrate their highly underdeveloped state, it is also their rather delicate appearance. The outer labia, for example, are structures that are only found in most other female primates while they are still in the womb, where they serve to cover and protect the developing sex organs. By the time a female monkey or ape is born, these outer labia have largely disappeared, leaving the clitoris fully exposed. Once they have done their job protecting the growing sex organs, they are reabsorbed. In human females, however, far from being reabsorbed, the labia remain throughout adulthood. And, as a consequence, a woman's genital opening, as well as her clitoris, stays somewhat enclosed throughout her life.

So what of the bonobo female, where does she fit into the scheme of things? Does the position and structure of her vulva resemble that of a female chimpanzee or a female human? The answer, as would be predicted by her partially infantised state, is that it is halfway between the two. Just like humans, bonobo females retain their outer labia into adulthood, while female chimpanzees lose these almost entirely as they mature. A female bonobo's vulva is also located far more between her legs than a female chimpanzee's, making the missionary position just as comfortable for her as being mated from the rear. In contrast, chimpanzees have never been seen to mate front-to-front – except, that is, for a single instance when a male and female were in adjoining cages. Far from simply 'being capable' of mating in the missionary position, it seems that female bonobos actually prefer to mate this way round, rather than from behind. This is indicated by the fact that they most often solicit sexual attention by lying on their backs with their legs apart, and sometimes insist on changing to this position even after a male has started to mate from behind. One possible reason for this is that the forward movement of the bonobo's vulva has made the clitoris far more exposed and forward-facing than in the chimpanzee. It is extremely likely, therefore, that front-to-front mating provides maximal stimulation for her.

Female bonobos reveal their infantile sexuality in other ways. Bonobos females are keen to have sex during most of their sexual cycle, rather than simply around the time when they are ovulating. In this regard, they behave like adolescent, rather than adult, chimpanzees. They have longer sexual cycles, also like adolescent chimpanzees, and return to sexual behaviour far sooner after giving birth. The result of these differences is that whereas chimpanzee females are sexually receptive for less than 5 per cent of their adult lives, bonobo females are receptive for as much as 50 per cent of the time. They are, in other words, roughly halfway between chimpanzees and humans, since human females are sexually receptive for all their adult lives. Once again, the bonobos illustrate how infantising creates a 'package' of changes – from a rounded head, to a flat face, long legs, and now front-facing female genitalia.

There is one final feature of women's genitalia that deserves particular mention because of its infantile origins. Completely hidden from view, it prompted Robert Smith of the University of Arizona and an expert on animal sexuality to declare that it 'is one of the great unsolved mysteries of human anatomy. I know of no plausible hypothesis for any physiological function it may serve, and I know of no other organ in the animal kingdom [that] evolved inevitably to be injured.' It is also the source of countless rituals and even murders all over the world. This curious, yet highly symbolic, structure is a delicate membrane that frequently blocks the opening of a girl's vagina, called the 'hymen'.

Named after the Greek goddess of marriage, the hymen is usually intact before a young woman has had her first experience of sexual intercourse. Moreover, since it frequently bleeds when it is first ruptured by a penis, it is frequently used as a virginity test in a wide range of different cultures. If there is blood on the sheet after the first marriage night, then the wife unquestionably merits the status of a virgin bride. However, if there is no blood, then her prior virginity is cast into doubt. Examples of an intact hymen being used as a test for virginity stretch back thousands of years. The 'tokens of virginity' mentioned in the Bible almost certainly refer to the bride's blood-stained nightclothes after her first night with her husband. Brueghel the Younger's painting of a village wedding feast clearly shows a bloodstained sheet hanging from the wall behind the rows of party guests. More recently, Eastern European immigrants to the United States have been known to send bloodstained sheets back to their families across the Atlantic to legitimise their marriages. In modern Egypt, it is still known for a bridegroom to wrap a piece of cloth around his index finger which he then inserts into the bride's vagina.

But of what relevance is this membrane to the infantile origin of adult women's genitalia? The answer is that this structure is little more than a remnant of an early stage in the development of female sex organs. It is a remnant of a plug of cells that disappears long before birth in all other monkeys and apes. Only in humans, just as with the labia, does the hymen persist well beyond birth. As with the plethora of other infantile characteristics that exist in our species, it has no purpose – instead, it just exists by virtue of our retention of infantile characteristics throughout our lives. If we want to find anything resembling the genitalia of an adult female human, all we have to do is take a look inside the womb of a chimpanzee, where an unborn female will still have a forward-facing vulva, outer labia and a hymen.

THE BONELESS WONDER

If chimpanzees and bonobos were suddenly accorded the status of 'partly human', they would pose something of a problem for those who wished to embrace them fully into the Jewish faith. The problem would arise when the primates were a mere eight days old, the time in their life when Jewish tradition requires that they participate in a ceremony called *brith milah* – the circumcision ceremony. The reason is that chimpanzees and bonobos, along with all other primates, do not have a foreskin. In fact, other primates lack of foreskin might even have been central to the origin of circumcision over four thousand years ago. According to zoologist Desmond Morris, the Egyptians, who were the first people to practise this form of mutilation, justified this custom at least partly on the basis that members of a baboon species that they considered sacred – the sacred or hamadryas baboon – were 'born circumcised'.

The human penis is unique in having a tiny hood that sits over its end, the foreskin. But to give the impression that human males have something that other primates do not have would be totally wrong.

The truth of the matter is that the human penis is simply arrested in an early stage of development. In other primates, the flap of skin that forms our species' foreskin carries on growing until it forms a luxuriant and ample fold of skin that encloses the whole penis. While a man's penis dangles vulnerably between his legs, the penises of other primates can retract into a large fold of skin, and thereby remain safe and snug when the male is not sexually excited.

Two words that succinctly describe the human penis are undoubtedly 'exposed' and 'floppy'. While our highly exposed penis is the result of our infantile condition, does this also explain why it is so limp? Do all primates have equally pliable penises? The answer is emphatically that they do not – and the reason is because, as most other male primates mature, their penis acquires a 'penis bone' that provides a certain amount of rigidity to the otherwise floppy organ. The penis bone is an unusual bone in as much as it is not joined on to any other bone in the body, but rather sits at the end of the penis completely unattached to anything else. The record for the longest penis bone is held by the stumptail macaque, a Far Eastern monkey and the King of Penis Bones, which has a bone measuring 6cm lodged firmly in the end of its penis. For those who derive pleasure from useless information, a curious feature of this particular bone is that it always resides in the left-hand side of the glans at the tip of the penis, and this explains why most monkeys and apes have lopsided tips to their penises – it is because the left side has a bone in it. Humans, however, are extremely unusual among the primates, and unique among the apes, in never developing this structure.

Of great significance, however, is the fact that in these other primates, the penis bone is largely absent in immature individuals. It is a structure that develops only as males become fully mature. Along with our highly diminutive foreskin, our missing penis bone provides further evidence of how our genitals remain in an infantile condition for the whole of our lives. It should be mentioned that, with characteristic human ingenuity, a small number of humans have partially managed to overcome the evolutionary loss of a penis bone. I am not referring to the unusual sexual practice of inserting a rod of ice, created by freezing water in a drinking straw, into the penis, but rather to the traditional insertion of a metal rod, called an *ampallang*, into the penis by the Dayak tribe of Borneo.

Making up for an evolutionary loss is also relevant when it comes to the human penis's lack of 'adornment'. In many other primate species the penis develops abrasive spines or bumps, as the male matures. As well as helping to prevent the penis from slipping out of the female, these spines are also likely to provide extra stimulation for both male and female. One account of the chimpanzee's penis describes it as being 'beset with numerous white horny spicules of yellowish-brown colour sharply contrasting with the rosy-pink of the intervening territory.' The human penis, in contrast, retains its smooth immature skin for the whole of its life – that is, unless it belongs to a man who is intent on making up for our immature genital deficiencies by surgical means. In a number of countries in South-East Asia it has been fashionable to insert various objects under the skin of the penis to make the shaft feel rough and bumpy, including bells, stones, jewels, ivory, gold, pearls, balls and shells. One visitor to this region made the following observations regarding penis implants:

> The bells ranged up to the size of a small chicken egg, and
> apparently were always made of metal, with a 'grain of sand',
> 'dried adder's tongue', or other object inside. The better ones

were 'gilded and made with great cunning'. As many as a dozen might be inserted. Kings might remove one of theirs to bestow it on a person deserving great honour.

In the Philippines, young girls can be seen selling plastic airgun pellets, called *bulitas*, to young men who then take these to back-street surgeons to have them implanted under the skin of their penis. In Japan too, the Yakuza, or Japanese mafia, still implant pearls beneath the skin of the penis to give it a nodular appearance. It is reputedly a tradition among imprisoned Yakuza that one pearl is allowed for every year spent in jail. As with the surgeons who operate on the young Filipino men, the Yakuza simply make a small cut in the skin covering the shaft of the penis, insert the pearl or other object, and then bandage up the wound until it heals. The result is a penis that looks not only bizarre, but somewhat diseased, since it appears at first sight to be covered with infectious pustules. As for the men's sexual partners, women seem to disagree as to whether the nodular penises are more sexually desirable than the natural sort, although a traveller to a part of India where a similar practice occurred at the time, did report that women invariably chose to have sex with a man with a 'titillating member'.

The human penis, then, provides a perfect example of our infantile state. It is so immature that it goes through its entire life without developing a pouch to keep it safe, a bone to keep it rigid, or spines to provide extra stimulation. While it is entirely in keeping with the overall condition of our infantile body, the immature condition of the human penis apparently renders it rather less efficient than it could be in stimulating the female. Men tend to assume that women 'naturally' take longer to reach orgasm than themselves. This is simply not true. According to the famous Kinsey reports on human sexuality, women can reach orgasm within just under four minutes by masturbating. In fact, 45 per cent of women can reach orgasm in between one and three minutes. Most men similarly need between two and four minutes to reach orgasm through masturbation. Perhaps most telling was the finding that, among women who masturbate, 95 per cent manage to have orgasms. Once a penis gets involved, however, everything changes. Firstly, during heterosexual lovemaking, the time taken for a woman to reach orgasm increases massively to between ten and twenty minutes. Secondly, Kinsey reported that among women who masturbate before marriage, over 50 per cent fail to reach orgasm regularly during the early years of marriage. Although there are undoubtedly psychological contributing factors to this change, it does appear that the human penis is not particularly efficient in arousing the female. In the words of so many school reports, the human penis's performance is 'satisfactory, but could do better'. Perhaps it would help if it matured just a little.

A WORLD OF DIFFERENCE

The human body, therefore, stands – not crouches, or stoops – as a monument to our infantile evolution. Every one of our species' unique features appears to be, in some way, linked to our retention of infant characters into adulthood. But there is a final twist to this story. It is that each of us has been infantised to a different degree. No two people are exactly alike, and different people will possess a slightly different combination of infantile characteristics. Features that are particularly infantile include a flat face, a large domed skull and high forehead, small ears, large lips, delicate bones, relatively smooth and hairless skin, and a low density of sweat glands – all these being characteristics that are associated with immaturity in our own as well as other primate species.

However, the most impressive variation is that which occurs between different races. Across the world, many of the distinguishing features of the human races have arisen through the persistence of infantile features to differing degrees. For instance, those characteristics in which Negroid Africans are particularly infantile include their flat noses, small ears, enlarged lips, long eyelashes, narrow joints, long legs and sparse hair on their skin. On the other hand, southern African Bushmen retain particularly infantile bulbous heads, small faces and persistent eye folds, while Caucasians retain the infantile condition of lacking pigmentation in the skin as well as having a moderately rounded head and high forehead. However, there is one group of people who lie at the furthest extreme of virtually every infantile feature that can be identified in our species. In fact, members of this group have been so greatly infantised that even some of their internal organs remain in an immature state for the whole of their lives. This group is the Mongoloids, and it includes most of the peoples of East Asia, such as the Chinese and Japanese, as well as the Inuits.

Adult Mongoloids' heads retain all the trademarks of infancy – including a rounded braincase, broad skull, small jaw, flat face, protuberant eyes, flat nose and high forehead. Among Japanese men in particular, the reduction in the growth of the upper and lower jaws has been so great that, for many, there is only just enough room in their mouths for all their adult teeth. The result is that many Japanese men have extremely crowded and rather badly fitting upper and lower teeth – a further example of how our species' infantisation has brought with it a variety of bizarre and potentially harmful side effects. Mongoloids' characteristic eye shape, with its 'epicanthic' fold of skin across the inside edge of the eye, provides another illustration of a structure that has distinctly infantile origins. Such a structure is present during early developmental stages of all humans while they are in the womb. In Mongoloids, however, this is retained not only into childhood but throughout adulthood. Any theories that try to explain the epicanthic fold on functional grounds, unavoidably suffer from the problem of being unable to explain why other races that inhabit similar habitats to those of the Mongoloids lack similar eye-folds. If, for example, it is suggested that the eye-fold evolved as a way of keeping sand out of the eyes in areas fraught with dust, sand and wind, then the question immediately arises of why is it absent from people who live in the the Sahara Desert, or any number of other places that experience tremendous sand or dust storms. No, just as with so many other features of the human species, the epicanthic fold exists simply because it is part of an extreme infantile 'package'.

As for the rest of the Mongoloid body, it is replete with further infantile features. Not only are the bones in Mongoloids' skulls unusually delicate, the same can be said of all the bones in their bodies – hence their typical light-framed stature. Perhaps the most obvious of all Mongoloids' infantile features is their relatively smooth and hairless bodies. The Chinese and Japanese, for instance, have very little body hair, as well as significantly fewer hairs per square centimetre on the top of their scalp. They also have far fewer sweat glands in their skin than other groups. Whereas Europeans and Africans have about 120 sweat glands in every square centimetre of skin, Mongoloids have a mere seventy. Indeed, in Japan, a strong body odour used to be sufficient in the early part of the twentieth century to warrant the sufferer being exempt from military service.

But it is in the details of Mongoloid anatomy that we find the most fascinating and conclusive evidence that their bodies, as a whole, are at the far end of the infantising scale. It is here that we find changes that almost certainly have no functional benefit whatsoever, but merely occurred as innocent passengers on the rest of the body's journey as it became more infantile. Among the various extreme infantile features – including the persistent juvenile muscles in the upper lip and cheek, the shallow articulation of the jaw, and the small size of the bony outcrop behind the ear – one above all others is particularly interesting. It is to do with the 'thymus gland', a two-lobed structure that lies in the upper part of the chest overlying the heart.

The thymus gland has been implicated in a number of vital processes that take place in young animals, such as the establishment of the immune system. However, it is also the source of a number of hormones, including a growth-promoting substance called promine – which is alternatively known as the 'Peter Pan Hormone' – a hormone that has the effect, when injected experimentally, of causing old animals to start behaving like young ones. For this reason, it has been suggested that promine is one of the central hormones in the regulation of the infantising process. So what has this got to do with the extreme infantising of the Mongoloid body? The answer is that in all races except for Mongoloids, the thymus gland is prominent at birth – when it is roughly the size of a newborn baby's fist – but largely disappears by the time adulthood is reached. It is only among Mongoloid people that this infantile gland persists right into adulthood and sometimes even into old age.

THE BIG QUESTION

This chapter has made what some might say is a biologically heretical proposition. It is that there were no advantages whatsoever to our species' new anatomical features when they first emerged. Far from being the result of precise evolutionary honing, and acute natural selection, the human species' most notable attributes arose 'by mistake' – they were merely dragged along as part of a major transformation that was occurring in our species. All the features that our ancestors possessed share the same infantile origins, and as such it is utterly pointless to search for independent explanations for the evolution of each and every one.

What is left, of course, is one lingering question. What could possibly have caused our ancestors to retreat into a permanently infantile condition? What was so good about the infantile version of our distant ancestors that made it supersede the adult version? This is the subject of the next chapter. For the moment, suffice it to say that the reason for our ancestors' transformation into permanent babies had nothing at all to do with the structure of their bodies, but instead something far more important for their survival six million years ago.

chapter two

babes in the wood

It is September 2010 and the beginning of the autumn television schedule. The broadcasters have exhausted their seemingly endless supply of reality game shows, with contestants having been filmed ad nauseam *as they attempt every conceivable challenge in every corner of the globe. Television executives are crying out for new ideas. Then out of the blue appears a proposal that has remained locked in a commissioning editor's drawer for over five years. It is an idea that, when it first arrived, was considered too macabre, too explosive and too controversial even to discuss. Nevertheless, because of its utterly shocking nature and undoubted sensationalist appeal, it was carefully filed away – waiting for just the right moment, should it ever occur, to present itself. If it could ever get past the broadcasting regulators, then it would unquestionably be the media sensation of the century. The proposal was labelled 'Pliocene Park', and it was the closest thing to an animal-versus-human gladiatorial battle since Roman times.*

The idea behind 'Pliocene Park' was simple and yet, potentially, breathtakingly brutal. A team comprising five to forty people – the precise number being decided by the team leader – was required to survive for forty days in Pliocene Park, a fenced game park in East Africa densely stocked with large carnivores, including lions, leopards, hyenas and crocodiles. Should all the team members survive for the full duration then they shared the prize of ten million US dollars. However, if any team member was killed, severely injured or refused to remain within the park until the end of the forty days, then the game would stop immediately and no prize money would be awarded. From the outset, every decision that the team leader took could make the difference between life and death. How many team members should he or she choose? Having more team members would almost certainly increase the chances of success and personal survival, but the pay-out would be split many more ways. Alternatively, having a very small number of team members would provide a large pay-out, but the risk of being overpowered by carnivores would become terrifyingly high. Perhaps even more importantly, what type of person should the leader choose to accompany him or her? At the two extremes, should their companions all be assertive, self-dependent and confident about going it alone? Or should they be willing to fall in line with the others and meekly do what they are told?

To the vast majority of viewers, the scenario would be fascinating for its human drama alone. But to anthropologists it would have a much deeper importance. The choice of a wooded East African game park was highly significant. So too was the need for the group to travel to find food packages that would be scattered around at random, as well as the rule that the contestants would have only their bare hands and what they could find around them for protection. Beneath its sensationalist façade, this programme was designed to recreate a scenario that is as *similar as possible to that of our early ancestors of about five million years ago – those who lived at the beginning of a geological epoch known as the Pliocene. Admittedly our earliest ancestors who lived at this time, with their slightly longer arms and more curved fingers, would have been better at climbing trees than us, but nevertheless their bones and fossilised footprints prove that they spent a considerable amount of time walking around on two legs. Although the game-show contestants would not know it, they were participating in the most daring scientific experiment of all time.*

Two years after the idea was first mooted in London, and with the reluctant consent of the broadcasting regulators, the game was finally ready to start. At 12 noon on Saturday 22 September 2012, over 500 million viewers in twenty-eight countries watched in silence as the three-metre-high gates were opened to allow the first team members to enter Pliocene Park. The atmosphere was especially intense because the first team leader to be selected, a 25-year-old man, had opted for the minimum team size permitted by the rules. A mere five people – three men and two women – walked silently through the gates with no means of defence other than their wits to keep them alive. With so few team members, each stood to win two million dollars if things went well. But as they all knew, it was a high-risk strategy and there was a definite possibility that one team member would never see the outside world again. As the gates closed, television cameras perched on watchtowers outside the enclosure relayed pictures of the five bodies as they nervously made their way along a narrow mud path that led into the woodland. Within minutes, they had all disappeared into the green tangle that was to be their prison for the next forty days. For the first time, nearly all the humans watching all around the planet were struck by just how terrifying nature could be.

The fictitious Pliocene Park is as close as it is possible to come to the birthplace of the earliest apemen – East African, open woodland, and filled with variety of carnivores and herbivores, many of which are remarkably similar to those which occurred two to five million years ago. The first elephants, for example, emerged in Africa about five million years ago from mastodon ancestors that had tusks that protruded from both their lower and upper jaws. Giraffes too were present at this time, as were early antelopes, rhinoceroses and horses. Although lions, leopards and cheetah did not evolve until one to two million years ago, the Pliocene contained many similar-sized carnivores, including the sabretooth cat *Dinofelis*, a predator resembling a modern jaguar, and the larger *Machairodus* which was more the size of a modern lion. Hyenas were also present during the Pliocene – their heyday, however, was slightly earlier, around 15 million years ago, and by five million years ago there was a mere handful of species remaining in Africa. Dogs, such as jackals and wild dogs, had yet to appear on the scene in Africa, although by five million years ago they had already made their way across the Bering landbridge from America where they originated, and spread across Europe and Asia.

As for the vegetation and climate at beginning of the Pliocene period, it was a time when Africa was suffering from dramatic cooling as well as major droughts. Vast expanses of luxuriant rainforest were progressively giving way to open dry woodland and ultimately to the wide open plains that are so typical of East Africa today. Only slightly earlier, around six million years ago, so severe were the droughts that the Mediterranean sea largely dried up, an event that involved sea levels falling an extraordinary 2 kilometres and that allowed many European species to enter Africa for the first time. It was a time of immense stress for the world's plant and animal life, during which 73 per cent of all mammal genera became extinct. Food, water and shelter provided by trees and bushes were all disappearing at an alarming rate. For tree-living primates, the situation was particularly dire. With the disappearance of the forest habitat, the drought was taking away the very essence of their livelihood. It was a time when our ancestors were at the very sharpest point of evolution – on the cusp of total annihilation. Ken Grimes sums up the situation, albeit with loose approximations for the names of the animals present at that time:

> What many accounts fail to emphasise is the hellish nature of this new environment. In the forest lurked sabretooths, leopards and giant carnivorous bears. More dangerous still was the ever-expanding savannah, patrolled by pack-hunting predators – giant flesh-shearing hyenas, blade-toothed dogs the size of wolves, lions and yet more sabretooths. There seems little doubt that these predators – already practised at hunting animals faster, larger and better-armed than hominids – would have found the conspicuous, puny and slow early humans a tempting meal.

Even today humans are highly desirable prey. Adrian Treves, a biologist at Conservation International in Washington DC, discovered that 393 people in Uganda were killed by carnivores over an eighty-year period that ended recently. He commented that 'these numbers undermine the long-standing assumption that large carnivores do not attack humans unless they are man-eaters as a result of disability or old age'. To this, psychologist Richard Coss adds, 'Once leopards and lions learn the ease of rasping flesh off human bones, they can become voracious killers, focusing almost exclusively on humans.

Such effects could have decimated ancestral human populations at night.' Indeed, a number of scientists now consider that in the past far too much emphasis has been placed on our early ancestors' ability to hunt, and far too little on the effect of them being hunted. Louise Barrett, a primatologist from Liverpool University, recently commented that 'It is surely more reasonable to imagine that early humans discovered weapons as a means of driving away predators, then adopted them for scavenging and then hunting ... I do think that the influence of predation on human evolution has been greatly underestimated.'

LIFE ON THE MENU

Whenever circumstances change, and a species suddenly finds itself at the top of the menu, it has no choice but to evolve – and fast. If it fails to do so, then the result is a short-lived feast for the predators, followed by one less species on earth. There are two options available to the hapless menu-item if it to survive. Either it can evolve new means of escape or defence – such as an increased body size, faster speed or weapons. Or it can change its behaviour – such as tuning its foraging time so that it misses the worst predators, grouping together or mounting counter-attacks on the predators.

So, where do our early ancestors – who, through force of climatic change, suddenly found themselves faced with predators of unimaginable ferocity – fit in to this scheme of things? How did they evolve in order to survive? Perhaps, however, this is the wrong question. Perhaps a more appropriate question to ask is which, of all the species that existed at that time, were best adapted to the new dry environment that superseded the luxuriant forests throughout much of

Africa? Of course, whichever question is correct, the result is much the same in the end – certain characteristics would have been successful in the new environment, and proliferated, while others would have disappeared.

But which characteristics would have best adapted our earliest ancestors to the new harsh environment? What type of ape would have fared the best? Was it one that was large enough to intimidate the predators, or fast enough to flee at great speed, or perhaps one that had massive weapons with which to defend itself? As it turns out, it was none of these. In fact, and as described in the last chapter, the situation could hardly have been more different, since what emerged was a frail species that tottered around on its back legs and had diminutive teeth. Instead of creating the incredible hulk, evolution had opted for one of the puniest apes of all time. Anatomically, we were pitifully equipped for life among the professional predators of the African bush. Clearly, on this occasion, evolution had gone for a behavioural solution to the problem. But what was it? Moreover, how was it achieved?

Without doubt, the best place to start looking for answers to these questions is among modern-day primates that live in environments where they are exposed to extreme levels of predation – particularly among those that live out in the open. As a matter of fact, there are relatively few primates that have managed to make the transition from the trees to the ground. However, when one compares the social behaviour of primates that do live in the open, with that of their relatives that live in the forests, a definite pattern emerges – those primates that live on the ground and in the savannah tend to live in groups that are significantly larger than those of their tree-living relatives. Among baboons, for example, those that live in the open African savannah travel in groups of up to two hundred individuals, while their forest-living relatives manage with groups of twenty to thirty. With their large numbers, mobs of savannah baboons can inflict serious injuries on predators such as leopards and cheetah. The same link between levels of predation and group size is seen in other primates. Macaques also form much smaller groups when they are living on predator-free islands than when living in areas where predators are abundant.

A similar pattern can be seen among different human populations today. The Kung Bushmen of the Kalahari in southern Africa, who live in a savannah habitat where predators abound, forage in groups of between twenty-five and forty, while forest-living tribes such as the Mbuti pygmies hunt in groups of two or three. Although small numbers of individuals are safe in the forest, they are totally untenable in an open habitat because of the threat from predators, where large number of individuals are required to gang together and use sticks or stones to intimidate or even kill predators. It is simply unthinkable that chimpanzees, baboons or humans should risk travelling across an open space in Africa in groups of only two or three. It would be lethal. For humans, as much as for baboons, the more exposed the habitat, and the greater the threat from predators, the larger the group needs to be. Life in the trees is like a summer picnic compared with the guerrilla warfare of life on the ground.

If any ape-like ancestors were to survive the devastating drying out of Africa five to seven million years ago, it had to be those that had the capacity to live in large groups. Inventing new means of defence or attack, such as sharp claws, or increasing the rate of locomotion by an order of magnitude was simply not possible given the bodies that our ancestors possessed. No, the only way for these puny apes to survive was through sheer force of numbers as well as through cohesive, mutually supportive cooperation. But how could this best be achieved? In an extraordinary case of scientific serendipity, the solution can be found in the forests of central Africa.

THE ANTISOCIAL RELATIVES

The scene that greeted the team leader and his four companions when they entered Pliocene Park was astonishingly similar to that of our earliest ancestors four million years earlier. It was an extensive dry woodland with wide-open spaces between each tree, containing life-threatening carnivores, the occasional herbivore, and themselves – a single species of ape that walked upright. What is more, it was an environment that the world's television audience could experience, albeit vicariously, by tuning into any one of the contestant's field of vision thanks to small cameras mounted on their heads. Of all the viewers none were more mesmerised by the scenario than the world's top anthropologists. To them, having five members in the team was highly significant, since this is the largest group size that chimpanzees usually travel in when they form foraging 'parties'. From a scientific perspective, the sixty-four-thousand-dollar question was, how successful would chimpanzee party-sizes be in an open woodland environment where there were large and dangerous gaps between the trees? In their natural forest habitat, chimpanzees spend 70 to 80 per cent of their time in trees, but this level of tree climbing was simply out of the question given the sparse vegetation of Pliocene Park. As the global audience watched the five contestants looking for a safe tree in which to spend their first night, many anthropologists hardly dared reflect on the established wisdom that linked high risks of predation to far larger group sizes than this.

With glistening teeth, glazed eyes and erect hair, a furious male chimpanzee can best be described as hell on legs. Beside himself with rage, he throws himself around the forest, screaming hysterically, beating and, if possible, destroying everything within his reach in an attempt to terrorise all around him. In this state, he epitomises everything that is aggressive, status-obsessed and belligerent about mature male primates as they constantly fight to elevate their position in the social hierarchy. Chimpanzees are clearly a far cry from the perfect animal when it comes to forming cohesive and highly cooperative groups.

Personally, I was first introduced to this extreme level of aggression when accompanying chimpanzee researchers in Liberia as they returned a chimpanzee to its colony after medical treatment. Rather than simply releasing the male chimpanzee, I was surprised to see that the researchers first placed him inside a metal cage within the enclosure. The reason, it was later explained, was that it was vital to check that the male had regained his strength fully before he was exposed to the rest of the troupe. In the past, the consequences of returning males too early had sometimes been horrific, with weak males being mercilessly ripped to shreds by other males who were keen to take advantage of this rare opportunity to elevate their position in the social hierarchy. Jane Goodall, the mother of chimpanzee behavioural research, recalls tale after tale of male aggression in her book *In the Shadow of Man*. In the following example, she describes the behaviour of Mike, an adult male who had just risen to the top position in her study troupe, and Goliath, who had previously held this position:

> By that time, however, his top-ranking status was assured,
> although it was fully another year before Mike himself seemed

to feel quite secure in his position. He continued to display very frequently and vigorously, and the lower-ranking chimps had increasing reason to fear him, since often he would attack a female or youngster viciously at the slightest provocation. In particular, as might be expected, a tense relation prevailed between Mike and the formerly dominant male, Goliath.

Goliath did not relinquish his position without a struggle. His displays also increased in frequency and vigor and he too became more aggressive. There was a time, toward the start of this battle for dominance when Hugo and I feared for Goliath's sanity. After attacking a couple of youngsters and charging back and forth dragging huge branches, he would sit, with hair on end, his sides heaving from exertion, a froth of saliva glistening at his half-open mouth, and a glint in his eyes that to us looked not far from madness. We actually had a soldered-mesh iron cage built in Kigoma and, when this had been set up in camp, we retreated inside when Goliath's temper was at its worst.

Even in captivity, when food is available in plenty, male chimpanzees constantly and relentlessly attack each other to raise or maintain their position in the dominance hierarchy. It is a feature of chimpanzee behaviour that has led to some extremely distressing scenes being witnessed in zoos across the world. One account that vividly illustrates male chimpanzee intolerance is that of Professor Frans de Waal who in the early 1980s was studying the largest colony of captive chimpanzees in the world, in Arnhem Zoo in the Netherlands. The colony consisted of three adult males and nine adult females, all of whom shared a large island and who were provided with an abundance of food to avoid unnecessary competition for this commodity at least. Even so, there was a considerable tension between the three males, and over the previous four years each of the males had managed to fight their way to the dominant position at one time or another. But nothing could have prepared Frans de Waal for what he found on the night of 12 September 1980 when, in his own words, 'the males' night cages turned red with blood'. Throughout July and August 1980, a chimpanzee called Luit had been dominant, but his reign was uneasy as he had never formed a stable coalition with either of the other two males, Nikkie or Yeröen. Although Nikkie would grovel to Luit, Yeröen barely even acknowledged Luit's status. It was a situation that, according to the observers, was bound to change sooner rather than later. And change it did when finally Nikkie and Yeröen combined forces to topple Luit. Although the fight on the night of 12 September was not actually observed, its result was horrendous. Luit lay among pools of blood, his body covered in deep wounds especially to his head, back, hands and feet. On the cage floor were scattered several of Luit's toes and fingernails – and both testicles. Luit died later that evening after three and a half hours in the operating theatre, leaving Nikkie in the top position.

It is little wonder, therefore, that chimpanzee cooperative groups, in the form of foraging parties, remain small and temporary. Male chimpanzees are simply far too belligerent and aggressive to work together in large numbers. As Frans de Waal commented, 'Male coalitions are instruments to achieve and maintain high status. There is little room for sympathy or antipathy in such opportunistic strategy ... Adult male chimpanzees seem to live in a hierarchical world with replaceable coalition partners and a single permanent goal: power.' Despite their antisocial tendencies, however, male chimpanzees have little choice but to frequently travel in groups, especially because of the need to protect their territory from neighbouring troupes by fighting off intruders. Temporary coalitions also make it far easier to catch small monkeys which are a prized delicacy.

As for female chimpanzees, although they are invariably less aggressive than males – 90 per cent of all aggressive incidences involved males – they are also far less likely to spend much time together or in parties. For the most part, they avoid contact with other females by dispersing themselves widely, and then restricting their movements to a relatively small area. Despite over a quarter of a century of intensive research, scientists are still unsure whether female foraging areas are in any way related to male territories – it might simply be a matter of chance whether a female's foraging area lies within one male troupe's territory or another's. For that matter, it has to be admitted that there are still many facets of chimpanzee social behaviour, especially those relating to the relationships between males and female, that still remain largely shrouded in mystery. What we can be sure of is that the only long-term associations that females form is with their offspring, who stay with them for up to ten years, and occasionally a very small number of other adult females.

For both sexes of chimpanzee, but females in particular, the levels of stress that would exist in a large group would be completely intolerable. In the case of males, it would be like a Molotov cocktail primed to go off at any moment. Such a high concentration of testosterone-fired fighting machines would be explosive. As it is, 40 per cent of all aggressive incidences occur when two foraging parties meet up and temporarily form a larger one – despite the fact that all the males involved belong to the same group. But stress can be highly detrimental to an animal's health in other ways. Among monkeys kept in zoos, it has been found that those at the bottom end of the pecking order are also those whose arteries fur up the most. In addition the lowest-ranking individuals not only have the highest levels of the stress hormone cortisol in their blood, but also the most depressed immune systems, meaning that they are more vulnerable to infections of every description.

For females, however, the cost of constant harassment can be even more severe and extend far beyond cuts, bites and grazes. This is because one of the most insidious effects of high levels of stress is to switch off the reproductive system. It is well known that women in high-stress jobs frequently find it difficult to conceive, and that lowering stress levels can solve the problem. Similarly, it is common to hear of a couple who, after years of trying in vain to have a baby, finally decide to adopt – and then, within months, the woman becomes pregnant. Furthermore, Thai and Danish researchers have found that women who work more than seventy hours a week in stressful conditions take significantly longer to become pregnant for the first time than women who work fewer hours. So too for primates that live in groups, in which the effects of long-term stress can be devastating on their reproductive success. In gelada baboons, for example, low-ranking females who are harassed a mere couple of times a day consistently have fewer babies than high-ranking females. In marmosets, tiny monkeys that live in South America, stress is so effective as a contraceptive, that so long as daughters live with their slightly bossy mothers, they fail to become sexually mature at all. Only when they leave home and occupy their own territory do the daughters undergo puberty. There are also early indications that the fertility of males too might be affected by stress. The first found that the sperm count of medical students plummets dramatically during exams, while the second reported that men who run over sixty miles per week had significantly lower sperm counts than normal.

The conclusion is straightforward. For male and female primates to live in the large groups that are essential for living on the ground, some mechanism needs to exist to reduce the inordinately high levels of aggression and stress found in chimpanzees.

ALL YOU NEED IS LOVE

At 4.05 a.m. the screams were clearly heard by camera teams positioned eight hundred metres away just outside the enclosure. Inside the enclosure, up an acacia tree, it was scene of pure terror and confusion as the whole team fumbled in the dark to climb ever higher. The team had been found by a pride of lions that roared beneath the trees. The two lowest contestants had managed to snap off branches and were beating these frantically against everything in their reach including, inadvertently, each other. For all five, their bodies were so consumed by fear that they shook uncontrollably and breathed in erratic shallow gasps. It was a primeval fear, the like of which none of the contestants had experienced even in their worst nightmare.

As the grey light of dawn penetrated the woods, the team realised the stark reality of their situation. Five feeble primates, with diminutive teeth, no claws and pathetically little speed, versus four adult lionesses, each weighing twice as much as an adult man, equipped with razor-sharp claws and a mouthful of massive teeth. The team had not even made it through twenty-four hours before they signalled their withdrawal from the game. The anthropologists' predictions had been proved correct. Regardless of the courage of each individual, small parties were simply incapable of defending themselves against predators in the semi-

open conditions of dry woodland. Even if the contestants had possessed the superior climbing skills of chimpanzees or our four-million-year-old ancestors, the trees were far too spread out to be able to climb from one to another.

To find the solution to the problem of how to make chimpanzees more sociable, it is time to return to little Mafuca – the darling of Amsterdam Zoo in the early 1900s. Being a bonobo, Mafuca is superficially very similar in appearance to a chimpanzee – however, when it comes to social behaviour, he could not be more different. Despite being the merest stone's throw away from the chimpanzee – genetically speaking – the bonobo has miraculously lost nearly all the aggressive and antisocial behaviour that so dominates chimpanzee society. Unlike chimpanzees, bonobos live in a peaceful bliss – in place of chimpanzee bickering and fighting for dominance comes bonobo mass 'love-ins'. To bonobos, aggression of the vicious skin-ripping type that exists in chimpanzees is virtually unknown. Any antagonism that does occur is invariably limited to chasing and screaming, and even after such encounters there is very little evidence of hierarchy, with the males usually engaging in conciliatory behaviour such as taking it in turns to mount each other, or rubbing their scrotums together while standing back to back. In this species, tension is characteristically resolved through sex – male with male, male with female, female with female, and even adult with infant. There is no creature on earth to which the phrase 'make love, not war' is more appropriate.

But more important still is the fact that bonobos generally live and travel in large groups. During the day, they move around in parties that are far larger than those of chimpanzees. At the longest established research centre at Wamba in Zaire, the average party size for bonobos is close to twenty, and parties of thirty or fifty are not uncommon. In contrast, chimpanzee parties usually contain a mere three to five individuals. Furthermore, whereas chimpanzee foraging parties are very much male-dominated, bonobo parties contain a roughly equal mixture of males and females. During the night, too, bonobos display their sociable nature by meeting up in mass sleeping parties. While chimpanzees sleep alone or with whoever they happen to be travelling with at the time, bonobos call each other during the late afternoon to encourage as many parties as possible to join up for the night.

In total contrast with chimpanzee society, in which males spend nearly all their time socialising with other males, bonobo males prefer the company of females to that of other males. It is almost certainly this lack of political manoeuvring between bonobo males, that has led to a fundamental difference between bonobo and chimpanzee societies. In chimpanzee societies, the dominant male is unquestionably Number One – he is treated with great caution, if not respect, by every other member of the troupe, including all the females. In fact, by the time that they reach adulthood, all the males are dominant to the females. But not so for bonobos. Because it was considered so unlikely, the suggestion that female bonobos might actually be dominant to males was dismissed for many years – that was, until a series of experiments confirmed it beyond a shadow of doubt. In one of these, primatologist Amy Parish compared the behaviour of bonobos and chimpanzees when identical groups – comprising one male and two females – were placed in an enclosure that contained a single pot of honey. Predictably, when chimpanzees were used, the male made a charging display through the enclosure and claimed everything for himself. However, when bonobos were used, the tables were turned. This time, the females approached the honey together, engaged in a session of mutual sexual contact, and then sat feeding side by side, taking it in turns to feed. Despite making a number of protestations in the form of charging displays, the male was totally ignored by the females who refused to budge. Such a finding tallies entirely with observations of bonobos in the wild, where rich sources of food such as large fruits on the forest floor are controlled entirely by females. While the females feed, the males have little choice but to charge around making futile displays, or to sneak close and attempt to steal scraps from infants – all of whom have free access to the food. It is an almost unbelievably far cry from the male supremacy of chimpanzee society.

The peaceful relationships that form the foundation of bonobo societies can, to a large extent, be traced back to the earliest relationships between mothers and their offspring. So strong is the dependency that sons have on their mothers, that males usually continue to follow their mothers around the forest even when they reach full adulthood. In bonobo society, the mother rules, and the sons retain their subservience to their mother for the whole of their life. Unlike chimpanzee males, bonobo males fail to develop the belligerent, dominance-obsessed, and incidentally rather misogynistic tendencies of adult male chimpanzees. In his book, *Demonic Males*, Richard Wrangham impressively describes the change that occurs in male chimpanzees as they take on the mantle of maturity.

At Gombe, when a male chimpanzee reaches adolescence he initiates a startling sequence of behaviours – a ritual, virtually – that has no equivalent among females … He enters the world of adult males by being systematically brutal toward each female in turn (when adult males aren't close enough to take sides) until he has dominated all of them... In a typical interaction, he might charge at the female, hit her, kick her, pull her off balance, jump on to her huddled and screaming form, slap her, lift her and slam her to the ground, and charge off again ... Nor does it stop there. In subsequent years, the male will often attack females without apparent provocation and with similar ferocity.

So what is it that underlies the fundamental difference between chimpanzee and bonobo social behaviour? Why can bonobos exist in large sociable groups, while chimpanzees squabble, fight and find it so difficult to tolerate the presence of more than a handful of other individuals? For that matter, could the same mechanism that makes bonobos so cooperative also have been responsible for allowing our ancestors to form the large supportive groups that would have been so vital for our earliest ancestors to survive on the ground?

The clues that can reveal the earliest stages of our species' evolution are now staring us in the face. Firstly, we have already discovered that bonobos, with their upright slender bodies and rounded heads, have highly infantised bodies. Rather than developing the heavy and well-fortified bodies of chimpanzees, bonobos retain many infantile anatomical characteristics for the whole of their lives. Secondly, it now appears that bonobo males fail to develop beyond the immature phase of being highly dependent on their mother for the whole of their lives – thus explaining why they follow their mothers around even when they are fully adult. Highly significantly, they also fail to develop the mature aggressive behaviours of chimpanzees as they head towards adulthood. As a result, bonobo males remain submissive to females.

The conclusion is as unavoidable as it is clear-cut. Not only are bonobos' bodies infantised, but also their behaviour – they not only look like grown-up baby chimps, but they behave like them too. By extending infantile and non-aggressive behaviour throughout their lives, bonobos successfully manage to live in far larger and infinitely more peaceful groups than their mature relatives, the chimpanzees.

As to whether our own ancestors went down the same infantising route as bonobos – to achieve the large group sizes that are so important for life on the ground – all we have to do is remind ourselves of the extraordinary similarity between our earliest ancestors' bodies and those of bonobos today. The reason for the astonishing similarity between bonobos and our earliest ground-dwelling ancestors is finally revealed. As bonobos demonstrate so impressively, infantising leads not only to an infantile body but also to the high levels of sociability that would have enabled our ancestors to move out of the trees.

ANATOMY MEETS BEHAVIOUR ...

Suddenly, everything starts to make sense. The apes that thrived in the dry woodlands of Africa around five million years ago, survived by virtue of their infantile tolerance of living in large groups and their ability to defend themselves from predators through sheer force of numbers. This explains why scientists have failed to come up with even one convincing explanation for so many of our species' unique anatomical features. It was not infantile bodies that were being selected for our early ancestors, but rather infantile behaviour. It was the latter that was so important for our survival on the ground, and our peculiar infantile bodies, with all their maladaptive components, just followed as part of the infantile package.

This process fits perfectly with what scientists have been saying for well over a century. Charles Darwin, for example, noted that 'if man goes on selecting, and thus augmenting, any peculiarity, he will almost certainly modify unintentionally other parts of the structure, owing to the mysterious laws of correlation'. In other words, if there is selection for one character, many others can be altered in the process. If an animal's behaviour changes from one generation to the next, this inevitably involves changes in body chemistry, which will, in turn, be highly likely to have knock-on affects on the rest of the animal's biology. What is more, precisely the association that is demonstrated by bonobos – that is, between infantile looks and infantile behaviour – has been described in a variety of other animals. Many adult sheep and dogs, for example, not only look but also behave very much like the young of their ancestors.

The strong link between infantile appearance and behaviour clearly points towards the futility of attempts to look for 'reasons' for the evolution of every feature when so many of these will have come as part of an overall package. In the case of our own species, so advantageous was it for infantile behaviours to be extended into adulthood – with all the cooperative and supportive consequences that this had – that changes to the rest of our bodies were, initially at least, a small price to pay.

Of course, walking on two legs does not make sense for a woodland ape – it simply makes them slow and bad at escaping predators. Similarly, it is extremely hard to see how the loss of large teeth for defence or the acquisition of a flat face could possibly have benefited our ancestors. The reason for these inexplicable and mysterious transformations is that they were merely incidental side effects of evolution selecting especially infantile individuals because of their particularly sociable and cooperative behaviour. The only apes that could survive in the new dry habitat were those that were able to live in large close-knit groups. Indeed, it was not just safety in numbers that was important, since disharmonious squabbling bands, however large, make far from effective and safe groups. As Adrian Treves has pointed out, infighting groups are noisy, conspicuous and less vigilant for predators while 'calmer, less conflict-ridden social organisations are safer'. Infantising, with its effect of simultaneously reducing tension, conflict and aggressive political manoeuvring – just as we see in bonobos today – provided the perfect adaptation to life on the ground for our ancestors by allowing them to form large cohesive and cooperative groups. What we are finding today, in the form of the occasional fossilised bone, is merely the incidental physical evidence of such behaviour.

Finally, with regard to the four-legged apes that are living in Africa today – namely chimpanzees, bonobos and gorillas – there are two possible explanations for how they arose. The first, more conventional, possibility is that they evolved from apes that survived the Pliocene drought in small pockets of remaining forest. The second possibility is that they evolved from the same two-legged ancestor that we did. Although the concept that all of Africa's apes are derived from the same upright ancestor as us might seem radical, it is supported by at least one astonishing fact. This is that, of all the ape fossils that have been discovered in Africa – that date to within the last six million years – every single one belongs to an upright ape. In other words, there is absolutely no trace of any four-legged ape having survived the Pliocene drought. If so, humans, chimpanzees, bonobos and gorillas could well be the descendents of a single upright ape ancestor – with humans being the only species to fully retain an upright posture, while bonobos have partially returned to a four-legged posture, and chimpanzees and gorillas have fully reverted.

MOB CULTURE

Within a week, because of the need to fill the now empty television schedules, the next group was ready to enter Pliocene Park. The programme commissioners in London were praying that this group would last longer than the first. No one had expected the first group to give up so quickly, but there again, none of those working on the project had experienced the appalling horror of being hunted by a wild animal.

The second group of contestants could not have been more different from the first. Having opted for the full complement of forty people, the group's leader – this time a 35-year-old businesswoman – had decided to adopt totally the opposite strategy to that of the first group. Rather than meekly skulking from tree to tree, she had chosen her group on the basis that they would stick together and jointly terrorise the forest, including all the predators, into submission. As soon as the huge metal gates to Pliocene Park were opened, the group sprang into action to provide themselves with weapons. Tightly packed and shouting at the tops of their voices they rushed towards the nearest trees and began to break off branches that would act as clubs and spears. Once armed, they started to play the most extraordinary and, from the television viewers point of view, the most unexpected game: they went off in search of the predators.

Two hours later the group found what they were looking for – a pride of lions. Three female and one male lion lay on their backs, legs stretched out and fast asleep, under a shrub at the edge of a water hole – oblivious of being watched by forty petrified, yet resolute, humans. For the contestants' female team leader, all was going perfectly to plan. If anything, she was surprised by just how well it was going. She had been dreading the first night if they had not managed to find any predators by that time. Still at a distance of one hundred metres, some members of the group put down their sticks and picked up as many stones as they could carry. Others armed themselves with stones in one hand while keeping hold of their sticks in the other. A few carried thin flaps of wood in each hand. All were doing exactly as they had been told. All were terrified to their core.

All forty people stood perfectly still, hardly daring even to breathe, waiting for the signal. With a piercing scream from their team leader, the whole group sprang into action in compact unison. More like one huge organism than forty separate individuals, the group emitted a hair-raising cacophony of screams, shrieks and shouts – the like of which neither the woodland nor any of its inhabitants had ever experienced before. As they ran straight towards the lions, adrenalin coursing through their veins, the individuals that together formed this cacopho-nous organism were by now numb to reality. The experience was more like that of a giant and complex nightmare in which all they could do was rely on a primitive autopilot, that had laid dormant in their genome for thousands of years, to carry them safely back into reality. It was as if, any minute, they would wake up to find themselves safe and sound in their English bedrooms.

As they ran and screamed their way towards the shrub that sheltered the lions, some members of the group threw rocks, while others waved sticks. The few carrying flat pieces of wood beat these together to add bangs and cracks to the already deafening riot. By the time they were fully up to speed, however, this intense experience was for the contestants' ears only. Within a fraction of a second of realising the monstrous nature of the creature that was heading towards them, with its vast dimensions and ghastly emanations, the lions had leapt up and raced off in a cloud of dust. As the forty humans arrived at the target shrub, the lions' absence signalled a victory that they could not yet appreciate. All they could do was slump down with the fatigue of those who had run hundreds of kilometres rather than a handful of metres. But for the group as a whole, it was a magnificent achievement, a rite of passage. Within the first few hours of its collective life, the group had already earned the right to call itself – so long as it remained and acted as a single organism – master of Pliocene Park.

The emergence of the earliest human ancestors around six million years ago would have heralded the world's first 'supergangs' – large, highly cohesive groups of apes that terrorised the African plains. While small squabbling groups of adults were perfectly viable in the forest, they were horrendously exposed to predation in more open environments. On the other hand, the juveniles possessed certain qualities that ideally suited them to a life on the ground – it was a time to retreat back from adulthood. From then on, as the fossils show, the only apes that could inhabit the dry woods and scrublands of Africa were those who remained immature for the whole of their lives. It was certainly not time to be consumed with status, bickering and infighting, but rather time to be infantile, cocky and brash. Out went individual aggression, and in came unruly yob culture. It was mayhem on the African plains where big, bold and boisterous groups of infantile apes bombarded predators with rocks and sticks. Whereas Africa's predators could previously eat the spoils of their hunts in peace, now they had to contend with groups of hooligan apes that roamed the plains and scrubland, driving them from their food like gatecrashers at a party.

The evolutionary pressure against these infantile apes reverting back to the dominance-oriented and violence-prone behaviour of their ancestors would have been immense, since doing so would have caused groups to split up and make individuals far more vulnerable to predators. Within each group, aggressive, antisocial individuals would soon become ostracised and be forced to brave it alone. So long as they all cooperated and did not bicker too much, they were safe and successful. Once it had evolved, the cooperative society of the eternal child was so powerful that any attempt to escape back to maturity was impossible.

As well as being less aggressive and more cooperative, immature apes have other qualities that would have been of value to our early ground-dwelling ancestors. There is good evidence, for example, that younger apes make the best communicators. An experiment that tested the likelihood and ability of different chimpanzees to inform others of where food had been hidden in a large enclosure concluded that 'the most dramatic and humanoid-looking signals were made by the most infantile and least efficient leaders, and they decreased markedly as the animals gained experience at leading other chimpanzees'. But, of course, one of the most noticeable qualities of immature animals, and especially primates, is their inquisitiveness and playful nature. At a time when our ancestors' regular food plants were dwindling, and as they were forced to find new sources of food plants, a high degree of curiosity that lasted throughout adulthood would have been a huge asset. By extending juvenile playfulness into adulthood our ancestors also had considerably more time to hone and develop their technical skills. While children love making train sets, adults love making incredibly complicated and skilfully constructed train sets – as my sixty-year-old father-in-law is in the process of proving much to the frustration of his neglected wife. Rather than having their playful experiments with sticks and stones limited to their early years, our ancestors could now improve their creative skills throughout the whole of their lives.

Japanese macaques – primates that frequently appear in wildlife films as they sit in hot springs in the snowy mountains of northern Japan – provide a wonderful example of how youthful experimentation can help adapt a primate to a new situation. In the early 1950s, a colony of Japanese macaques that lived on the edge of a small bay in Japan became the subject of an experiment to see how well they coped with a new type of food. They were provided with sweet potatoes which, unlike the macaques' normal diet of fruit, were covered in gritty sand that made them difficult to eat. In September 1953, an eighteen-month-old female called Imo was the first to solve the problem. She took her sweet potato to the water and washed the sand off before eating it. Over the next few months other members of the colony imitated Imo's behaviour and also washed their potatoes until, five years later, seventeen individuals in all were washing their food. However, of those that learned this new trick, a mere two were adults while fifteen were aged between two and seven years old. Eighty per cent of the adults in the troupe failed to pick up the new technique, despite observing it every day.

Although large-group formation was the major motivating force driving the infantising of the apes, the total package – such as other coincidental benefits including an extended period of playfulness and learning – was a massive success. At first sight, members of this wunderkind species might have seemed slightly laughable, as they tottered around on their two back legs with pathetically little armoury, natural weapons, or even speed. But this new species was on its way to becoming far greater than just a collection of individuals. Instead, like a deadly Portuguese-man-of-war jellyfish, or a swarm of African bees, each group of upright apes was behaving increasingly like a single invincible organism. It would be a few million years more before the full potential of this infantile species was finally realised, but nevertheless the ball had inexorably started to roll.

As soon as they had decided on a sleeping site, which was a small glade with a towering boulder on one side, the members of the team organised a rota for being on watch. Since no clocks or wristwatches were allowed in Pliocene Park, the changeover times would be a bit hit-and-miss, but none of the team had any doubt as to how important it was that at least a few people were awake at all times. If any noises were heard, it was agreed that those on watch should immediately shout the alarm, whereupon everybody would get up, grab their sticks and form a tight group – all the while shouting, screaming and making as much of a racket as possible. Of course, with no fires allowed and in the pitch black of the night, there was the chance that there might be some confusion – but given all the noise there should be little chance of anybody failing to locate the others.

During their first night in Pliocene Park, the team members found it virtually impossible to sleep. It was not that they were excessively fright-ened, as those in the first group had been, but rather that they were still experiencing the after-effects of the adrenalin rush that had accompa-nied their assault on the lions. As the night progressed, however, excite-ment gradually gave way to exhaustion and by the early hours of the morning, all those who were not on watch were fast asleep. Thankfully, and as it turned out, unusually, this first night went by without any disturbances and all the team members slept soundly on their home-made mattresses of leaves and branches.

Each of the following thirty-nine days and nights had its own incident, but not one caused the group to even consider giving up. Their strategy was simple and extremely effective – if they encountered any predators, be they lions, leopards or even crocodiles, the group would immediately close ranks and attack. It worked perfectly every time, with the hapless carnivores retreating at great speed under a barrage of flying rocks and stones. After a while, the group began to split into two during the day in an attempt to locate the food that was hidden each day by Pliocene Park staff. But the group never split below twenty individuals, and at the end of the day they either met up at the main camp, or called to each other to make sure that they were reunited by nightfall.

At 12 noon on Wednesday 7 November, each of the forty contestants picked up a cheque for $250,000 as they walked out of Pliocene Park. For them, it had been a magnificent success. Once they had established that they could terrorise the park's predators by staying in large groups and attacking them en masse, there was very little else to be worried about. For the television companies, on the other hand, they were left with a major problem. Although Pliocene Park had achieved the highest ratings ever, where could they go from here? Now the world knew that success in the park was virtually guaranteed if the contestants formed tight rowdy groups, there was very little point repeating the exercise.

chapter three

a kindergarten world

Cairo well deserves the epithet 'mother of the mega-cities'. Within this one sprawling mass of humanity, 70 million people dwell in every available space, from high-rise apartment blocks, to rooftops, pavements, underground passageways and even among the gravestones and mausoleums of ancient cemeteries. Not even termites live in anything like this type of number they reach a maximum of roughly five million individuals in any one colony. Yet, apart from petty crimes such as pickpocketing and theft, Cairo has a remarkably low violent crime rate. For the vast majority of the time, its inhabitants quietly get on with their own lives, finding a multitude of highly imaginative ways to earn an income. Moreover, Cairo is far from unique. Throughout the world, humans live in vast cities, in which aggression is the huge exception to an otherwise peaceful rule. The success of humans in avoiding aggressive encounters with members of the same species is simply astonishing. The same evolutionary mechanism that enabled our ancestors to form large cooperative groups, and thereby survive on the woodland floor alongside vicious predators, has rendered us superbly adapted to living in close confinement with millions of others.

Before proceeding any further, it is important to justify the claim that human societies are virtually free from aggression – particularly since the media constantly bombards us with tales of violence from around the world. For a start, it is important to make a clear distinction between inter-personal aggression, of the type that is spectacularly rare in human societies, and warfare. It is not that war does not involve aggression – because of course it does – but rather because it is a highly distinctive behaviour that has no parallel in the animal kingdom. Human warfare, in the vast majority of cases, is not concerned with the total annihilation of 'rival' groups – as it would be in, say, ants or wolves that are fighting over a particular territory – but instead is about forcing or persuading others to share a common allegiance, either to a spiritual or corporeal leader. To sidestep humans' natural aversion towards aggression towards members of the same species, leaders do all they can to avoid their soldiers perceiving the opposing army as being made up of like-driven individuals – by creating different costumes, languages, and other distinctive cultural emblems – lest this should introduce weakness in the form of sympathy. However, most importantly of all, the very existence of armies in the first place – in which tens of thousands of individuals are willing to work together for a common cause – bears testimony to a staggering level of cooperation and lack of aggression within our groups. Whereas chimpanzee hunting groups are limited to 5 or so, before they disintegrate through aggression, humans groups can contain well in excess of 500,000 individuals and nevertheless contain extraordinary levels of cohesion. So, warfare aside, let us let us consider the evidence regarding human aggression in everyday life.

Of all the urban situations in the world, New York has one of the worst reputations for violence, especially homicide. So where better to start looking for evidence of our species' peace-loving and supremely tolerant disposition? Using the Federal Bureau of Investigation's crime figures for New York, it is possible to ascertain the total number of reported incidents in one year that involve violent behaviour – a category that includes murder and non-negligent manslaughter, forcible rape, robbery and aggravated assault. The total figure given for such incidences during the year 2000 is 75,745. Clearly, the real number of violent assaults is going to be higher than the reported number – however, the latter should at least provide a reasonable indication of the number of serious assaults. Armed with this figure, it is then possible to work out roughly how often an inhabitant of this reputedly savage city will experience serious aggression.

To do this, however, it is necessary to estimate the total number of interactions that each New Yorker has per day that involve more than a passing word or polite acknowledgement. How many people, in other words, do we spend more than a few minutes talking to each day? For the purposes of this exercise, let us assume that each of us has on average ten such interactions. So, given that the population of New York in the year 2000 was 7.7 million, this means a total of 14,000 million interactions per year – of which, according to the FBI, 75,745 are violent. In other words, one in every 186,644 interactions involves serious aggression. Again taking an average of ten interactions per day, this translates to each New York inhabitant being personally subjected to one serious aggressive incident in every fifty-one years. What is more, although the way that statistics are compiled varies from one country to the next, the frequency of aggresive incidents is roughly comparable in other countries that publish similar data. For example, in my own home city of Bristol in England with a population of 400,000, the figure comes out at one violent incident per person every thiry-two years, while in Oxford it is once every thirty-three years, and in England and Wales combined it is once every thirty-six years. If children under sixteen are excluded from the population figures for England and Wales combined, the frequency of violent incidents per person is slightly higher – but still a mere once every twenty-nine years.

It therefore seems entirely valid to claim that human groups possess astonishingly low levels of aggression. By failing to develop mature characteristics, we have been tamed. We are essentially a domesticated version of our distant wild ancestors. What is more, for compelling evidence of the power of infantising to transform wild creatures into those that not only possesses extraordinarily docile qualities but also thrive in close confinement, we need look no further than our own sitting rooms and backyards.

IN OUR IMAGE

Around 14,000 years ago, humans began the long journey that was to lead to them playing God. Although they had coped reasonably well on their own up to this point, they began to realise that an assistant would be of immense value – a supercreature capable of working alongside them that could complement their own talents. It was an aspiration of which Victor Frankenstein would have been proud. From then on, in villages all over the world, many generations passed as humans created the perfect companion to suit their particular needs. The difficulty, however, was getting this creature's qualities just right. Some needed to have superhuman powers of speed, some needed to be fearless sentinels, while others needed extraordinarily keen senses. But, above all, none of these creatures should, under any circumstances, turn on their masters – they needed to be wholly subservient to humans, immensely cooperative and do exactly what they were told. The latter, needless to say, was a quality that Victor Frankenstein spectacularly failed to achieve.

Fortunately for all of us, those who were involved in moulding this creature did not turn to body parts exhumed from graves, or to massive electrical voltages plunging into vats of amniotic fluid – both in true Frankenstein style – but rather to adapting another animal that had long been showing an interest in the human way of life. The wolf. The problem for our ancestors was simply how to turn this wild and often aggressive creature into one that would fit in well with human society, be sociable, loyal, hard-working, cooperative and reliable – in fact, one that in as many ways as possible displayed the characteristics of the ideal human being. So what did our ancestors do to transform this wild animal into one that possessed this ideal collection of human-like qualities?

Darcy Morey, an archaeologist from the University of Tennessee, specialises in the origins of domestic species and has investigated the unique physical characters of early domestic dogs. After taking measurements from sixty-five adult prehistoric, but domesticated, dog skulls – procured from museums in America and Europe – Morey compared the dimensions of these with the skulls of wild species of dog, including wolves. The avoidance of modern dog breeds was clearly important because of the plethora of 'fancy' characters, such as highly extended or compressed muzzles, that have recently been bred into them. Morey's findings tell something of a familiar story.

Early dogs conveniently exhibit consistent morphological [structural] changes when compared with wolves. Briefly … dogs became smaller overall, and the length of the snout became proportionately reduced. The result was a smaller animal with a shorter face, a steeply rising forehead and proportionally wider cranial dimensions. This general pattern suggests that adult animals retained juvenile characteristics …

When I plotted width-to-total-skull length proportions, I saw some interesting differences. Adult dogs are distinct in these dimensions from all the adult wild canids. But adult dogs do resemble one wild canid group: juvenile wolves.

Morey also noted that the earliest dogs frequently had crowded teeth, sometimes overlapping each other, in jaws that are not really big enough to accommodate them efficiently. Another distinctive infantile feature of domestic dogs is their small canine teeth. At the other end of the body, domestic dogs also rarely, if ever, develop a functional 'supracaudal', or tail gland, a structure that serves an important role in the social life of mature wild canids. In coyotes, wolves and foxes, the tail gland is highly visible roughly a third of the way down the tail, either as a round patch or a dark streak running down the tail, and is the source of a musk-like secretion. In the few instances where a dark area does develop on the tail of a domestic dog, little or no glandular tissue exists beneath this marking, confirming the incomplete development of this feature.

So how does the behaviour of domestic dogs compare with that of their wolf ancestors? Does this in any way mirror dogs' infantile physical characteristics? The answer is that adult domestic dogs display a dazzling array of wolf puppy behaviours. When we go out, our canine companions sit patiently at home awaiting our return – like wolf puppies at the den waiting for the rest of the pack to return with food. Then, when we arrive back home, they rush up to us, jump up, and try to lick our face, nose and mouth – just as wolf puppies do when they attempt to get their parents to regurgitate food. Finally, we bend down to give them attention, and they roll over in an act of total infantile submission – sometimes even urinating, again much like wolf puppies when their mother returns to the den. It is all a far cry from the behaviour of a nervous and snarling wolf that skulks in the corner of its cage when humans approach. Instead, domestic dogs retain puppy-like playful and submissive behaviour for the whole of their lives, which they direct towards their human 'parents'. Darcy Morey describes the situation:

Many adult dogs not only appear juvenile, they also act juvenile. They display a sort of behavioural paedomorphosis [infantising]. Dogs routinely solicit attention, play, grovel, whine, bark profusely and otherwise exhibit behaviour that wolves more or less outgrow as they mature. Biologist Raymond Coppinger and linguist Mark Feinstein describe dogs as 'stuck in adolescence'. They also make the important point that the essence of tameness is the submissive, solicitous behaviour style of juveniles.

A delightful illustration of how domestic dogs typically fail to develop beyond an infantile wolf stage comes from an experiment that Harry and Martha Frank conducted using an adult Alaskan malamute – one of the oldest Arctic sled dogs – and a wolf puppy. The particular situation arose because, as wolves and other wild dogs grow up, their infantile dependency relationships soon give way to adult dominance/subordinate relationships. With domestic dogs, however, such a transition rarely happens – which, of course, is highly desirable in the home where a permanent dependency relationship is far preferable to one based on competing for top status. In the case of the malamute and wolf puppy, this difference resulted in the young wolf puppy exercising its status behaviour in the presence of an adult domestic dog that had little idea of what such behaviour meant.

> The inability of our adult Malamute to comprehend this interplay of dominance and submission behaviours has been a source of constant tension in our group … One of the male wolf pups formed an especially strong attachment to the old dog and tried to subordinate himself at every opportunity. Lacking any cooperation from the dog, this sometimes forced him into contortions that became more absurd as he grew older and taller. At about 5 months of age, he was trying eagerly to elicit some response to his submission behaviours when the Malamute yawned. The pup managed to thrust his entire head in the dog's mouth, and when the dog found his jaws clamped round a wolf pup, he appeared as astonished as the pup was delighted.

Just as humans can be viewed as infantised chimpanzees, so domestic dogs can be viewed as infantised wolves. As well as retaining many infantile behaviours, the adult dog also stands apart from all its wild relatives in retaining the infantile yelp into adulthood. In wild species of dog, such as the wolf and coyote, yelps are emitted only by puppies for the purpose of attracting the attention of the mother. As for barking, one of the most characteristic and often most irritating features of domestic dogs, this sound too is rarely made by the adults of wild species, although it is common in juveniles. When adult wolves do occasionally 'bark', it is far softer and more muffled than the deafening machine-gun-like outbursts of our domestic sentinels. A particularly vivid illustration of the difference between the vocal behaviour of wolves and domestic dogs can be provided by placing both in unfamiliar surroundings. Whereas wolves become completely silent, domestic dogs whine, yelp and bark to attract the attention of their parent substitute – which is invariably their human companion.

Humans have employed exactly the same process that moulded our own species to make the wolf the perfect human companion – they infantised it. Both the appearance and the behaviour of the domestic dog illustrate how our ancestors must have systematically favoured, or perhaps initially tolerated, those dogs who remained as puppy-like as possible throughout their lives. For canine society to merge with human society, it too had to undergo the same infantile transformation to avoid the high levels of aggression, the rigid territoriality and the inflexibility that accompanied the transition to maturity in their wild ancestors. Fortunately for us, wolves were somewhat primed for such a development since they were already partially infantised to enable them to work together in cooperative and cohesive packs. Compared with other wild dogs, the wolf retains considerably more infantile behaviours into adulthood – these being referred to as 'derived infantile actions'. The theory is that such behaviour patterns – for example, a strong tendency to follow superior animals – helps in the integration and coordination of pack activity. In just the same way that our own ancestors adapted rapidly to their new environment – that is, out of the trees – by merely extending certain infantile features into adulthood, so dogs have become adapted to their new human environment in remarkably short time and through exactly the same process. Accordingly, Robert Wayne, of the Zoological Society of London, has studied the DNA sequences of modern canids and found that dogs differ from wolves merely in their developmental timing and growth rates. What this very clearly illustrates is how, given sufficient environmental pressure, species such as dogs and humans can become transformed through the infantising mechanism within an incredibly short time – in the case of dogs, a mere 14,000 years or so.

Starting with extremely nervous and aggressive predators, generations of breeders have managed to create the perfect companion – an 'honorary human' – that has the capacity to ease our lives in a myriad different ways. From hunting dogs, to rescue dogs, shepherd dogs, sniffer dogs, egg-collecting dogs, fishing dogs, racing dogs, sled dogs, truffle dogs, police dogs, dogs that see for us, dogs that hear for us, dogs that provide companionship, and even dogs that act as hot-water bottles for us, we have created the perfect partner – infantile companions that are keen to submit to their 'parent's' instructions. It seems that nothing is too much for these animals that have been carefully moulded, above all else, to dote on us. We have, in other words, managed what Victor Frankenstein so miserably failed to do – constructed a creature in our own [idealised] image. Inevitably, there is still an element of competition and squabbling between dogs – as exists between puppies in a litter – however, this is very different from the all-out naked aggression and status-obsession that predominates wolf society. It should also be mentioned that, in much the same way that humans have recently and intentionally reintroduced a variety of 'wolf-like' looks into certain breeds, so they have reintroduced extremely aggressive and confrontational behaviours. It must be emphasised that these are not behaviours that belonged to the original domestic dog – of the type that worked alongside our ancestors – but instead those of modern recursive monsters that have been created by an equally dysfunctional human society.

In dogs as in humans, however, all is not necessarily plain sailing in an infantile world. On the one hand, mature behaviour patterns occasionally re-emerge causing serious antisocial problems, especially when this involves a struggle for dominance and raised levels of aggression. On the other hand, some dogs have become so infantised that they are almost completely neurotic. Because many humans find dogs with extreme infantile qualities highly attractive, various breeds have ended up almost unbelievably retarded in terms of their behavioural development. Many have gone far beyond the designed compliant and cooperative partner, and have become more like excessively demanding babies. A recent study found that King Charles spaniels, for example, have the behavioural repertoire of a wolf puppy that is less than twenty days old, while Norfolk terriers, Shetland sheepdogs and French bulldogs fare little better on the wolf developmental scale. Furthermore, in dogs and humans alike, infantising seems to carry with it certain psychological risks and potential problems. Pet psychologists, for example, recognise the 'perpetual puppy' syndrome in dogs that have been over-pampered in the early part of their life. The consequences of such an upbringing for an already infantile species can be as peculiar as they are varied. In both humans and dogs that are excessively mollycoddled during their early

years, separation from their carer can lead to severe depression
including submissive withdrawal. Alternatively, a fear of separation
and lack of attention can lead to anxiety-reducing compulsive eating,
or 'sympathy' illnesses in which the sufferer feigns health problems
for the sole purpose of getting attention. Sympathy lameness, paralysis of the legs and joint pain, have all been described in dogs as well
as in humans as attention-seeking devices. The following account
describes just such a condition:

A veterinarian in Des Moines, Iowa, was presented with a
Pekingese which was unable to use its hind legs. He was unable
to make an immediate diagnosis and the pet was left in his hospital for further examination. Within half an hour the dog was
standing up on all fours and wagging its tail. He called the
owners to come and collect it, which they did, but within one
hour they had returned, the dog again in a state of collapse. The
veterinary surgeon then inquired further into the home background of this dog to find that the woman had just had a baby
and although the dog had been her centre of attention for a
number of years, her attention was now focused almost exclusively on the newborn child. This was too much for the overde-
pendent dog and it developed a classic conversion hysteria
involving paralysis of both hind legs. The treatment, which
immediately resolved the syndrome, was to place the dog in
another child's home where it lived happily ever after.

Despite its immense social benefits, infantising and the excessive
dependency it can create is undoubtedly at the core of a wide variety
of psychological problems and hang-ups regardless of species. This is

a subject that will be revisited later when our own species' infantile behaviours are explored in detail.

SPICE OF LIFE

Dogs are certainly not the only animal to have been domesticated by becoming infantile versions of their former selves – our farmyard animals have been rendered more docile through exactly the same process. All are characterised by an almost entire lack of the dominance, aggressiveness and territoriality of their mature ancestors. Take domestic pigs, for example. Descended from vicious and highly-territorial wild swine, living in herds of five to eight under the control of a tyrannical alpha male, the European domestic pig is a paragon of docility. Living comfortably in peaceful herds of many hundreds or even thousands, the domestic pig seldom shows any aggressive behaviour whatsoever towards its fellow inmates, even when it is kept under extremely confined conditions. But just as importantly, domestic animals frequently bear the physiques of the juvenile stages of their wild relatives. Domestic pigs, with their rounded bodies, short faces, small teeth and thin unpigmented skin, are far closer in appearance to huge bloated piglets than their wiry dark ancestors. In other domesticated species, infantile features similarly include reduced or absent horns, slender bones, a greater proportion of fat and muscle, a reduced thickness of the skin, as well as the short jaws and flatter faces of many traditional breeds.

Many of our present-day domestic animals are likely to have originated from wild species that were already partially infantised well before humans became involved. Just as in our own species, infantising appears to have occured in a number of other species as an adaptation to allow group sizes to become significantly larger. In the early 1960s, Valerius Geist, a zoologist from the University of Calgary, carried out a classic study of wild mountain sheep in North America. Of all the sheep that Geist observed, he considered the Rocky Mountain bighorn to be the most highly infantised. Rams belonging to this variety not only lack the magnificent neck ruff and beard that are typical of mature males in other species, they also fail to completely develop mature behaviours such as kicking subordinates with their front legs. Referring to the bighorn males as 'cowardly', Geist concluded that 'The mature individual resembles behaviourally the juvenile of the original parent population and not the adult'. But most importantly, the infantised bighorns are also by far the most gregarious of the wild mountain sheep that Geist studied. Behaving more like giant lambs, they live in groups that are roughly two to four times the size of other species. Such a variety would, of course, have been ideal for further domestication.

Domestic animals, like us, have been suspended in a state of permanent infancy to prevent them from developing mature and aggressive behaviours, and to allow them to live together in large numbers and high concentrations. We have surrounded ourselves by gaggles, herds, flocks and packs of infantile versions of wild species. What is more, if we carry on destroying the world's remote places and wild species at such a high rate for much longer, we will soon turn our planet into one large kindergarten. In place of noisy herds of wild animals, in which mature males constantly fight for dominance, will be peaceful pens of docile cows, sheep and pigs that meekly accept whatever living conditions are provided for them. One big happy infantile family.

As well as our infantile origins and tolerance of living in large groups, there is another quite unexpected similarity between humans and domesticated animals. It concerns the extraordinary amount of physical variation that exists in all domesticated species – including humans. Take a look at a wolf and what do you see? The answer is a wolf. Instead of what you *do* see, however, it is rather what you do *not* see that is so fascinating. What you do not see is a long-legged wolf, or a dwarf wolf, or one with a red coat with white splotches all over, or with floppy ears, or with a flat pug face. What you are guaranteed to see is simply a regular wolf. On the other hand, in domesticated species the sky is virtually the limit when it comes to the number and type of different varieties that exist – there being about a thousand different breeds of domestic dog, displaying a breathtaking variety of colours, shapes and sizes. A common assumption is that every tiny feature of every breed has emerged through painstaking selective breeding – from colour pattern, nose shape, hairlength, eye colour, ear shape, tail length, etc. But this is an absurd proposition. Although many of these features have been subsequently reinforced through selective breeding, it is their appearance in the first place that is so puzzling. In the case of my own family's dog, a Welsh springer spaniel, her red-and white coat has no doubt been modified by selective breeding so that it forms an attractive pattern, but this is a long way from explaining where the red colour came from in the first place. Moreover, the same extraordinary degree of variation that exists in dogs also exists in other domestic species – including, of course, humans that come in a wonderful variety of sizes, colours and shapes. Again, it is not so much what different cultures have done with this variety, in terms of favouring long legs, red hair, small female breasts or any other feature, but rather the existence of this variation in the first place that is so remarkable.

So what is behind all this variation? Is it simply a matter of predators normally culling such odd-looking individuals out of wild populations?

It appears not, and that levels of spontaneous variation are a hundred to a thousand times higher in domesticated species than in their wild ancestors. At 'normal' wild mutation rates, there is nowhere near enough time in the 14,000 years or so since the first dogs appeared to create all the breeds that exist today. Rather, it seems that a core feature of domestication, and infantising itself, is the destabilisation of the whole developmental process and, through this, the spontaneous appearance of a vast array of different varieties. Evidence for this comes not only from every veterinary surgery and farmyard, but from one of the most ambitious and lengthy biological experiments of all time.

Dmitry Belyaev is a colossus within the scientific community. His experiments are the scientific equivalent of the artworks of Christo – the artist who is best known for wrapping vast landscapes and national monuments in cloth. Born in a small village 350 kilometres north of Moscow, Belyaev studied genetics in Moscow before serving in the Second World War, first as a soldier and then as an officer in the Soviet Army. After the war, he returned to work in a genetics laboratory in Moscow until, in 1958, he moved to western Siberia, which is where he carried out his magnum opus – an experiment that was to carry on for forty years and involve over 45,000 animals. Belyaev was fascinated by the phenomenon of domestication which he considered to be one of the world's greatest and longest-running experiments. But of all the aspects of domestication, what Belyaev found particularly interesting was the fact that, regardless of which species was involved, exactly the same features turned up time and time again. Across the board, from cows to dogs, goats, and guinea pigs to cats, the effect of domestication was the same: it caused the loss of a seasonal pattern to reproduction; it led to giant and dwarf varieties; it gave rise to a huge variation in coat pattern, colour and hair texture. In addition, it consistently made the animals' ears go floppy and their tails go curly. Despite the apparent disparity between these changes, it was Belyaev's firm belief that one factor alone was responsible for all of them – the selection for tame and docile behaviour.

To test his theory and find out what lay behind all these changes, Belyaev decided to replay the domesticating process from beginning to end, using an animal that was similar to the dog but, as yet, completely wild. He chose the Russian silver fox. Starting in 1959, he systematically rated each fox on a scale of 1–3, according to how tame they were towards their human handlers. Category '3' foxes would flee from people or bite them when stroked or handled. Category '2' foxes allowed themselves to be handled but were not friendly towards people. Category '1' foxes were friendly to people and wagged their tails enthusiastically. In every generation, only the very tamest foxes were allowed to breed. After a mere six generations, all Belyaev's foxes were so tame that he had to create another category, which he called 1E. Foxes belonging to this generation were so eager to establish human contact that they whimpered to attract their handler's attention, and once in contact with humans they sniffed and licked them excitedly just like dogs.

The vital point to remember is that Belyaev was selecting his foxes using one criterion alone – their tameness. Nothing else mattered or was relevant in any way. Yet, like all other domesticated animals, the same bizarre features started to crop up virtually as soon as his experiment started. After eight generations many of the tame foxes were born piebald – unlike any of their wild ancestors, they lacked pigmentation in parts of their body and developed a variety of different patterns on their fur. Some also had curly tails, while others retained the floppy ears that are normally restricted to young wild foxes. Many also kept on making puppy-like sounds right into adulthood. After fifteen generations, unusual lengths of tails and legs appeared, as well as changes to the jaws and muzzles. In a mere handful of generations, such was the level of novelty that it was like walking out of Wall Street into the middle of the New Orleans Mardi Gras. Belyaev had not selected for any of these novel features, they had simply arisen spontaneously along with tameness.

However, Belyaev had a strong suspicion of what might link domestication with the emergence of physical novelties. What he suspected was that, by selecting for tame behaviour, he was actually selecting animals that had a different chemical balance in their body, and that this was bound to effect other aspects of the animal's biology. It is exactly the same phenomenon as that which caused so many physical changes in the human body during our evolution – select for a different behaviour, and the rest of the body changes at the same time. What is more, Belyaev was right. Along with the arrival of the characteristic domestic features, such as piebald coats and floppy ears, came a massive drop in the levels of stress hormones normally found in wild foxes. Simultaneously, there were also huge rises in the levels of the hormone 'serotonin' in domestic foxes' brains – this being a hormone that is thought to have an important role in inhibiting aggression as well as, incidentally, in shaping an animal's development from a very early stage.

The cat, or perhaps more appropriately, the dog, was out of the bag. What Belyaev proposed was that selecting for tame behaviour results in the severe disturbance of chemicals that control 'normal' development in the wild ancestor. Animals that fail to develop some of the more mature behaviour patterns of their ancestors do so because they have unusual levels of certain hormones in their body and this in turn causes other aspects of their biology, such as their growth and hair-covering, to differ too. Belyaev described this as the 'destabilising' effect that domestication has on the development of an animal, which results in 'a great increase in the range and rate of hereditary variation'. Moreover, because most mammals, including those that humans have domesticated, share similar developmental systems – in terms of the chemicals involved – this explains why domestication has caused similar physical anomalies to emerge in each of them.

The significance of his research findings for the evolution of human physical variation did not escape Belyaev's notice. Towards the end of his career, he described how our species too must have experienced the same conditions that his foxes faced in his experiments, in which there is selection for individuals that remain docile and sociable. Such conditions, Belyaev noted, are exactly those that give rise to a severe destabilising of a species' growth and development, and that cause a species to display a fantastic amount of variation.

'Variety's the very spice of life / That gives it all its flavour' – so wrote English poet William Cowper in the eighteenth century. Little did he know, however, that he had an extraordinary quirk of nature to thank for this gift. Were it not for the need for our ancestors to become domesticated to enable them to work cohesively together, instead of squabbling and bickering for dominance as did their ancestors, we would be a very dull and uniform species indeed.

BIGGER GROUPS, BIGGER BABIES

Before leaving the subject of domestication altogether, I just want to mention one last and intriguing possibility. Since infantising is strongly linked to a reduction in aggression between members of the same species, is it possible that this is the evolutionary key that opens the social door to animals throughout the animal kingdom? We have already seen how wolves are partially infantised, and how this contributes towards a greater degree of pack coordination and cohesiveness. We have also seen how dogs, as well as other domesticated animals, have become increasingly gregarious and tractable as they have been infantised. As will be discussed more fully at a later stage, dolphins and whales, with their complex social systems, also have many features that suggest that they are highly infantised versions of their cow-like ancestors. It has even been suggested that some of the great herding animals, such as African buffalo, manage to live in such large peaceful groups because of a failure to develop mature territorial and socially aggressive behaviours. Richard Estes, an authority on African mammals, regards the retention of juvenile non-territorial behaviour as the most successful social strategy among Africa's cow-family grazers, and suggests that 'the most socially advanced – i.e. the most gregarious – tend to achieve the highest population densities and to be ecologically dominant'. And finally, of course, humans are by far the most infantised of all the apes, as well as being by far the most social. But how far does this link between infantising and social behaviour stretch? What about the world's ultimately social animals – those that biologists call 'eusocial' – in which some individuals devote their whole lives to looking after the offspring of a single 'queen'? Has infantising got anything to do with the highly social behaviour of these species?

In fact the evolutionary scenario that could lead to a species becoming eusocial, through becoming more infantile, can apply to virtually any species – from mammals to insects. All it takes is for the offspring to take increasingly longer to mature and become independent until, finally, they spend so long in the place where they were born that they participate in looking after the next batch of offspring that their mother produces. The more infantised a species becomes, then the longer the offspring stay in the nest, and the greater the chances that they will participate in raising their younger brothers and sisters.

So let us take a brief tour of the zoo to explore whether the world's eusocial species are also the most infantile. First stop, the insect house. In one corner of the room, there is a small display containing a few adult and hundreds of newly-born cockroaches. In the other corner is a vast towering mud tower that contains over a million of the most social insects of all – termites. Looking into the heart of the termite mound, it is possible to catch fleeting glimpses of the tiny insects as they rush up and down passageways. But what a surprise. It appears that some of the baby cockroaches have escaped and have made their way into the termite display. The creatures inside the termite tower block look almost identical to the baby cockroaches on the other side of the room. Actually, nothing is wrong, and there is a simple explanation for the mistake. Termites can logically be seen as hugely infantised cockroaches, since there is a staggering resemblance between termite workers and newborn cockroaches. Various termites display a broad range of features that indicate their infantile origins, from their missing or tiny eyes, to their reduced or absent genitalia, shortened antennae and jaws with partially developed teeth. In the words of E.O. Wilson, insect biologist and authority on animal social behaviour, the workers of some termite species can most aptly be described as 'child labor'. It has even been suggested that all social insects share a similar infantile origin, as demonstrated by the fact that the workers of many species remain wingless.

As we continue our journey around the zoo, beyond the gregarious buffalo, and past the entrance to the giant and highly social mammal 'tadpoles' that inhabit the dolphinarium, we arrive at the doors of the underground mammal house. This is home to a living legend – the only eusocial mammal that exists – one that uniquely lives in tight-knit communities in which most individuals are destined never to have babies of their own, but instead spend all their time looking after those of the queen. Perhaps uniquely in biology, the existence of this particular animal was predicted many years before it was actually found.

Naked ape meet naked mole rat. Naturally occurring beneath the feet of elephants, lions and the other animals of the East African plains, the naked mole rat is one of nature's extreme oddballs. Functionally, it is like a drilling machine with its cutting points permanently exposed. Socially, it is like a termite, in having a queen and various different castes of workers. Visually, it is the closest thing that exists to a sausage – and at 10cm long it is even the same size. With its naked wrinkly skin and tiny undeveloped eyes that are used merely to sense air currents in the pitch blackness of its subterranean home, the naked mole rat is one of the few animals that looks almost identical regardless of whether it is a day or a year old. In other words, the adults display a virtually identical set of infantile characteristics as the babies. They fail to grow up. Shunting soil up and down tubes, chain gangs of workers tirelessly move soil from the soil face to the surface. Without large numbers of workers, the chances of digging sufficient tunnels to find the juicy plant tubers on which the colony depends for food would be untenably small. With naked mole rats, just as with termites, success and survival in this difficult environment has come about wholly through living in large highly-cooperative communities. Moreover, this extreme level of social behaviour is once again found in a species that simultaneously displays a wide variety of infantile features.

The link between infantising and highly social behaviour does indeed stand up. Animals such as humans, bonobos, domestic dogs, and a swathe of other domesticated species, not only possess infantile bodies and behaviour, they also form larger, more cohesive and peaceful groups than their 'wild' and non-infantised relatives. And now it seems that the world's most social animals of all – the eusocial animals – are also highly infantised versions of their less social relatives. Could it be that the only way of achieving a very complex society, comprising hundreds, thousands, or even millions of compliant individuals each performing their own prescribed task, is by reducing each to a permanent state of infancy? Might it even be the case that if our species becomes still further infantised, we will head towards the social organisation of termites in which most individuals fail to reproduce? In fact, this might not be so far from the truth, for reasons that will become clear in later chapters.

chapter four

original sin

Although infantising has undoubtedly had a major effect on a species' social behaviour, arguably its most fundamental effect has been on the relationship between the two sexes. In termites and naked mole rats, for instance, it has even gone so far as to render the majority of individuals incapable of reproducing altogether. So what effect has it had on our species? In this chapter we explore the origin of human sexual relationships. Have we always fallen in love? Are we naturally promiscuous or polygamous, and is monogamy forced on us by an overzealous and Procrustean society? If a romantic story is expected, involving devoted husbands and wives jointly looking after their angelic children, then prepare to be disappointed. Love may be one of our species' most endearing emotions, but it is also one that has an especially devious origin – involving sexual manipulation, female guile and the ruthless exploitation of our most infantile cravings.

THE TANGLED WEB

The Chinese emperor Fei-ti undoubtedly earned his position as one of the most oversexed men in the history of our species. Not only did he have 10,000 women in his harem, he was even taught from an early age to conserve semen so that he could fertilise two women per day. Fei-ti was a sperm factory. Indeed, to improve his reproductive success his court officials kept careful records of dates of menstruation to avoid the emperor wasting his precious seed. Fei-ti, moreover, does not stand alone in his prodigious sexual appetite. The Inca king Atahualpa, for example, kept 1,500 women in each of his palaces that spread across his kingdom, who were reserved for his sexual delectation alone. If any other man violated any of his women, then he, his wife, his children, his relatives, his servants, his fellow villagers and all his llamas would be executed without question. The Babylonian king Hammurabi also had thousands of female slaves at his sole command, while the Aztec ruler Montezuma had 4,000 concubines, and the Indian emperor Udayama had a record-breaking 16,000 consorts who were kept in apartments ringed by fire and guarded by eunuchs.

There are also more recent examples of people who have had a colossal number of sexual partners. The Belgian crime writer and creator of the Maigret character, Georges Simenon, claimed to have had sex with a different woman every fifty hours throughout his adult life, while the French actress Mlle Dubois named 16,527 in her diary of lovers which, if true, is the equivalent of having a different lover every night for forty-five years. Perhaps most impressively of all, King Ibn Saud of Saudi Arabia allegedly had sex with three different women every night for sixty-three years, making a total of 65,000 different sexual partners in a single lifetime.

On the other hand, throughout history, there is a plethora of examples of men and women who have become so emotionally attached to one individual that they would rather die than be subjected to life without their sweetheart. In the second century, for instance, Appian of Alexandria wrote:

At the beginning of the third century, B.C., Seleucus, one of Alexander's generals and among the ablest of his successors, married a woman named Stratonice. Antiochus, his son by a previous marriage, had the misfortune to fall in love with his new stepmother. Recognising the illicit character of this love, and the hopelessness of its consummation, Antiochus resolved not to show his feelings. Instead, he falls sick and strives his hardest to die.

Appian then goes on to recount how the celebrated Greek physician Erasistratos was summoned and how, after several days observing Antiochus, he concluded that the only solution was for Seleucus to divorce Stratonice, and for Antiochus to then marry his erstwhile stepmother.

More up to date, in 1979, Dorothy Tennov interviewed more than five hundred lovers, among whom was Ruth – a woman who exemplifies the type of person who is so besotted with a single other individual that the rest of the world might as well not exist:

How can I say that my seemingly interminable passion for Eric, a 15-year obsession, was reasonable? Consider the 30,000 hours – I actually calculated my estimation, and that's a conservative figure – I spent going over every word he said, every gesture, every letter he wrote, when I might have been reading, or learning a foreign language, or enjoying the company of others. Instead, I was caught in a merry-go-round of wondering how he felt, wishing he would call, anticipating our next time together, or endlessly searching my recollections of his behavior and my convoluted reconstruction of the possible reasons for his actions for the shreds of hope on which my madness fed.

So, where does this leave our species in terms of its 'typical' sexual behaviour? Is it ideally polygamous, as in the case of Fei-ti? Or does it crave promiscuity, as exemplified by George Simenon and the dissolute Giovanni Casanova – the archetypal Italian seducer of women? Or, finally, is the human species essentially monogamous, as illustrated by the fact that by the time Americans reach their late twenties, 70 per cent of men and 85 per cent of women have formed stable monogamous relationships? These are questions that have lingered in the minds of academics and laypersons alike for thousands of years as they have attempted to cut through the Gordian knot of human sexual variation. Despite the obscuring tangle of emotions there is, however, one evolutionary path that leads straight to the heart of our species' sexual behaviour, and its origin can be found in the sex lives of our distant ancestors.

PROMISCUOUS BEGINNINGS

To be brutally honest, science has really got only one way of guessing what the mating system of our extinct ancestors might have been, and this is based on the size difference between the males and females. The hackneyed rule is simply that species in which the males are considerably larger than the females are usually polygamous – the reason being, as Darwin first suggested in 1871, that large body size confers a definite advantage in contests between males for access to females. The larger the number of females that each male guards, then the more intense the competition, and the greater the size difference between males and females. In the monogamous gibbons and marmosets, therefore, the males and females are roughly the same size, while male gorillas that jealously protect harems of three and five females, weigh a massive 210 kilograms, which is twice that of the females.

So how large were our male and female ancestors of four million years ago, and what might this say about their mating system? The answer to the first question is, on the face of it, that the males were about twice as heavy as the females – and on the basis of this, the vast majority of anthropologists conclude that our earliest ancestors were strongly polygamous. I say 'on the face of it' because of a niggling concern that scientists have mistakenly mixed up the bones of more than one species, and arbitrarily labelled the bones from small species as belonging to females, and the bones from large species as being 'male'. The underlying problem is that it is very difficult to determine whether a particular bone came from a male or female. Moreover, we know that for the majority of our species' evolution there have been a number of upright ape species in existence at the same time. For example, two million years ago there were as many as fifteen different species of upright ape. But, despite this doubt, nearly every book on human evolution reaches the same conclusion – that, on the basis of the difference in size between the males and females, our earliest ancestors were polygamous. Males, so the story goes, were large because they competed with each other for exclusive sexual access to their harem of females. There is only one problem – even if our ancestral males were bigger than the females, there are other far more tenable explanations for why this was the case.

Imagine you are living in Africa four million years ago, and one of a tight-knit group of thirty upright apes that are cautiously making their way across the woodland floor. In this dry environment, teeming with bone-crunching predators, life is a constant struggle for food and survival. Imagine too, that a handful of males are jealously guarding all the females. As for the remaining males, they have no option but to make do with the occasional sneaky copulation behind the harem owners' backs. There is always tension between the 'haves' and 'have-nots', and fights for possession of the females are common. Then a sabretooth cat appears. What happens next? The answer is that there is utter mayhem and confusion. As harem owners attempt to guard their females despite the danger, some males rush off hoping that harem owners are killed so that they can claim their females, while other males rush after females that have become separated from the rest of the harem, and so on. The fundamental problem is that hugely unequal distribution of resources – in this case sexual partners – creates intense competition and infighting, and this makes it extremely difficult to operate as a cohesive team. For this reason, in species where large numbers of males are needed to repel predators, obsessive guarding of females is simply not tenable.

Hamadryas baboons in Ethiopia, primates that possess a harem mating system, illustrate the major obstacles that such a mating system puts in the way of male cooperation:

A male threat-gapes at an opponent during a band fight. The rival turns away, grasping his nearest female around the hip. This possessive frenzy is accompanied by a strong inhibition in the male to even look at another female … The reproductive career of a young Hamadryas male is successful if he obtains possession of a harem. His difficulty is that all reproductive females in a troop belong to some male who will fight any encroaching rival … When fully grown, male followers in one band attacked the ageing family leaders in their clan and took some of their females by force. Grown males of other clans joined and sometimes obtained females from the disrupted harem.

But there are other serious objections to our early ancestors having a polygamous social structure. The second relates to the movement of males and females into and out of the group. In nearly all primates that have a polygamous mating system, when the males reach sexual maturity they are chased out of the group by the dominant male to avoid competition. Young males are forced to chance their luck with females elsewhere. So what of our own species? How do we conform with this general rule? The answer is that in the vast majority of human societies around the world it is not the males but rather the females who emigrate from the group into which they were born to find a sexual partner.

Many traditional societies today still have ceremonies to mark the stage when a young woman moves to a neighbouring village to live with her new husband. Throughout history, there are numerous tales of female transfer – by foul means as well as fair. The Scandinavians, for example, often fought for and wedded their wives at sword-point, and women were frequently abducted on their way to the church to be married – this being the reason why marriages were generally cele-brated at night. Similarly, the role of the best man was traditionally to prevent the wife from being abducted immediately before the wedding vows have been made. Maoris and Fijian marauders, on the other hand, made more of a feast of female theft – by eating a few of the local men before abducting their wives. Perhaps the most extreme case of violence towards a future wife comes from an account in *Chamber's Journal* of 22 October 1864, following an anonymous writer's expedition to Australia, in which he describes a common way that an Aboriginal obtains a wife:

> He hovers round the encampment of some other blacks until he gets an opportunity of seizing one of their *leubras*, whom perhaps he has seen and admired when attending one of the grand corroborries. His mode of paying his address is simple and efficacious. With a blow of his nulla-nulla (war club), he stuns the object of his 'affections', and drags her insensible body away to some retired spot, whence, as soon as she recovers her senses, he brings her home to his own gunyah in triumph. Sometimes two join in an expedition for the same purpose ... Slowly and silently, they creep close enough to distinguish the figure of one of these *leubras*, then one of the intruders stretches out his spear, and inserts its barbed point among her thick flowing locks; turning the spear slowly round, some of her hair speedily becomes entangled with it; then, with a sudden jerk, she is aroused from her slumber and as her eyes open, she feels the sharp point of another weapon pressed against her throat. She neither faints nor screams; she knows well that the slightest attempt at escape or alarm will cause her instant death, so, like a sensible woman, she makes virtue of necessity, and rising silently, she follows her captors. They lead her away to a considerable

distance, tie her to a tree, and return to ensnare their other victims in like manner.

By whatever means, in the human species it is traditionally the females who move out of their birth group, not the males. The males stay right where they were born, thus making it very difficult for a few males to monopolise all the females. Very recently in our species' history, the situation has become more complex because of the ability of a few men to accumulate wealth – for example, in the form of cattle, land or money. However, because farming emerged a mere 10,000 years or so ago, for the vast majority of our hunter-gathering history, such resource accumulation would have been impossible. Indeed, among humans that still practise a hunter-gathering lifestyle, in which males join forces to hunt for food and protect the group from predators, polygamy is hardly known. For our distant ancestors, such would have been the need to protect their group against predators and possibly neighbouring groups, that the males formed close bonds and stayed in their birth group for life – just like chimpanzees today. So long as this happened, then polygamy would have been impossible.

Whereas male cooperation and female transfer make it extremely unlikely that our earliest ancestors were polygamous, what is entirely compatible with these essential features of our species is a promiscuous mating system. Not only is this the mating system of our closest relatives, the chimpanzees and bonobos, but scientists such as Jared Diamond consider that it is also the most likely mating system of the common ancestors of chimps and humans. As chimpanzees demonstrate so impressively, a highly effective way to achieve male cooperation – albeit on a small scale – is by allowing young males to stay in the group, and by allowing all of them to have at least some access to females. This certainly is not to say that there cannot be a male pecking order within a promiscuous mating system, especially when it comes to having sexual access to females, but this is a totally different situation from that in which a large number of males are ostracised by a few high-ranking ones who monopolise all the females. In chimpanzees, unlike polygamous species, there is no evidence for long-term relationships being formed between particular males and females, and some scientists have even questioned whether high-ranking males even have any advantage over low-ranking males in their mating success at all.

While it is so easy to interpret this promiscuous sexual arrangement, with pecking order on the side, as a masterly plan of males, it is the females who should in fact take most of the credit, since it is they who largely decide how many males mate with them. If the females form tight groups, for instance, then polygamy becomes an option since the males can guard many females at once. However, if the females spread themselves out, then guarding them all at once becomes a nightmare, and polygamy is virtually impossible. Alternatively, females can influence the mating system simply by making vehement protestations when certain males attempt to mate with them. This is delightfully illustrated by Jane Goodall's observations of a troupe of chimpanzees in Gombe, Tanzania:

> One young female, slightly older than Fifi, showed a marked objection to the advances of the aggressive male Humphrey. Gigi invariably had a large male retinue when she went pink – from the first time she had shown a true adult swelling it seemed that she had as much sexual appeal as Flo; but she simply could not bear Humphrey. When all the other males were satisfied,

there would be Humphrey, his hair on end, glaring at Gigi, shaking branches, hunching his shoulders, stamping with his foot on the ground, moving cautiously toward her. All the while Gigi would be screaming and moving away from him ... Fifi also avoided Humphrey, though she was less frightened of him than Gigi had been and merely walked calmly away when he began to court her. Once we watched Fifi walking and cantering around a tree no less than fifteen times with Humphrey pursuing her, all his hair bristling. He could easily have caught her; instead, he finally charged away in a wild display of frustration, hurling huge rocks, stamping his feet, and vanishing completely from the scene. Similarly, other females avoid other males.

Females are likewise very much in the driving seat of society's sexual arrangements in savannah baboons. For many years, scientists looked at the imposing size and teeth of male savannah baboons and concluded that this meant that they could 'insist' on mating with whichever females they were allowed, by rank, to mate with. In retrospect, this conclusion actually says more about the scientists who were studying baboons than the baboons themselves. Going by what we now know about baboon sexual behaviour, it says that the scientists were all men and that there was a long history of male scientists seeing species with larger males than females as being totally male-dominated. As Shirley Strum, an American scientist who turned the traditional scientific view of baboon behaviour totally on its head, commented:

There was no mistaking its compelling message [coming from male scientists]: males were the building blocks and the cement of the group. They were the focus and the power. They were the structure and the stability, the essence and the most valuable part. Male baboons were fighting machines, with powerful muscles, thick mantles of hair and razor-sharp canines. With such equipment, males competed with one another for whatever good things there were: food, females, and a place to sit.

What Shirley went on to do was to prove that quite the opposite was actually true – that males had to be almost obsequiously nice to a female before she would consider letting him mate, something that usually involved endless hours of grooming her and her baby – one that, incidentally, was very likely to have been fathered by another male – and that males had to work incredibly hard to persuade a female to mate with him. To her utter astonishment, especially given all that she had been told by her male supervisors before she left America, Shirley also discovered that in her group at least a male's rank seemed to have very little effect on increasing his chances of mating with a female, in fact it sometimes even decreased his chances. She could not fail to be moved by the social equality that existed between the males and females in this savannah primate, and felt sure that her findings were of immense relevance to the emergence of our own species. 'If my initial findings were accurate, the implications would extend beyond baboons to interpretations of the evolution of our own human behaviour ... aggression, male dominance and male monopoly of the political arena are not necessary aspects of the lifestyle of the earliest humans.'

For a wide variety of reasons, therefore, polygamy seems to be incompatible with what we can deduce about the lifestyle of our early ancestors, while promiscuity is far more likely. But surely we have ignored the central pillar of the argument for our ancestors being polygamous. What of the link between big males and polygamy – surely this is the ultimate indicator of the mating system of our ancestors, with big males resulting from intense competition over possession of a harem? How do we explain the fact that our four-million-year-old male ancestors may have been as much as twice as heavy as the females?

For the answer, we need look no further than East Africa's savannah baboons – a species that not only lives in a habitat that is virtually identical to that of our early ancestors, but that also has a society in which males do not fight over exclusive access to females, and in which the males are nevertheless twice as heavy as the females. So what explains the large size of males in this species? The reason is purely and simply to do with living on the ground. By comparing plains-living baboons with their close forest-living relatives, three main changes emerge as being associated with a move into the open. Firstly, and as already mentioned, group size increases. Secondly, a move out of the trees is linked with an increase in overall body size. Savannah baboon males are about three times heavier than their tree-dwelling relatives, the macaques and mangabeys, the most likely reason being that larger size provides vital defence against predators in more open habitats. Likewise, our early ancestors also underwent a considerable increase in size as they moved from the trees to the ground – from 10–20kg for our tree-dwelling ancestors to over 45kg in our upright terrestrial male ancestors of four million years ago. Thirdly, in ground-dwelling species, males are relatively much larger than females compared with tree-living species. While this can partly be explained by the larger overall size of ground species – since one effect of increasing size is to exaggerate any size differences between the sexes – it is also the case that larger males are better able to protect themselves and other members of their group, in particular the young which may be their own offspring, against attack by predators.

But there is one further explanation for a marked size difference between males and females that is all too often overlooked by zoologists and anthropologists. It concerns the obvious fact that a size difference between the sexes is as likely to be caused by females becoming smaller, as by males becoming larger. Anthropologist Robert Martin has devised an ingenious method of testing whether 'small females' or 'large males' are responsible for sex differences in size across the primate group. Working from the premise that certain structures – in particular, the brain and teeth – are relatively resistant to size change, regardless of what happens to the rest of the body, Martin proposed that if females' bodies have become smaller then their brains and teeth should be relatively big, whereas if males have become larger, then their brains and teeth should be relatively small. And the results? These fly in the face of previous assumptions that sex differences in size have resulted almost entirely from males getting larger. Martin found that there was a far greater tendency – among simian primates – for females to have relatively big brains and teeth than for males to have relatively small brains and teeth. His conclusion was that at least as much of the size difference that exists between the sexes in primates can be attributed to females getting smaller as to males getting larger.

So, if our own species similarly followed this same general trend, why should our female ancestors have stayed smaller than expected when they moved down from the trees to the ground? Evidence from other species has led Martin to suggest that small female size may be linked to particularly high rates of predation – the rationale being that it is beneficial for females to have babies early in life, even though they are still small, if the chances are high that they will be killed by a predator sooner rather than later. Once a female begins to channel her energy into having young, this essentially brings her growth phase to an end. Furthermore, small female size is also associated with the young being born relatively large for their age – this way, they will be more likely to survive if they are abandoned by the mother, either because she is killed or because she is having more offspring. Overall, the more dangerous or unpredictable an environment is, the less time a female potentially has to reproduce, and the more important it is that she produces independent young early in her life.

At last a tenable explanation for the marked size difference between males and females in our ancestors has emerged. Without any reference to polygamy, a perfectly logical explanation is that males were larger than females to provide better protection against predators – such as sabretooth cats and carnivorous bears – while females remained small because of the advantages of reproducing early in life in an environment where death was possibly just around the corner. Having been fully liberated from the specious argument that big males can only be explained by polygamy, it means that there is now no reason to think that our ancestors were anything but promiscuous.

In the light of this, it is now possible to ask if our species is still fundamentally promiscuous? Did the increased sexual freedom of the 1960s, for example, reveal that when society relaxes its rules, promiscuity emerges as our 'natural' mating system?

THE UBIQUITOUS URGE

In terms of coming up with a devious scheme to satisfy a desire for wanton sex, the alleged escapades of Cyrus Spragg take some beating. Having first established a nudist colony in Michigan – which fell victim to cold weather and irate objections from local inhabitants – Spragg and his ex-Mormon followers moved to Cairo, Illinois, where he ordered the construction of an ark similar to that of Noah's in the Book of Genesis. Despite Spragg's insistence that there was going to be a deluge and the only means of survival would be the ark, the modest rains that year meant that Spragg yet again had to fold his operation and move on. His next stop was New Jerusalem, and this time he embarked on an ambitious project to build a temple and ecclesiastical palace in the style of King Solomon during his reign over the Israelites. Once completed, Spragg proceeded to inform his followers that he was going to enter the temple for good, but continue to direct them as an 'Eternal and Invisible Presence'. Just one other thing, he wanted a different virgin to be brought to him every night – in complete darkness since no one was allowed to look upon the face of the Invisible Presence – with the promise that one of them would become the modern Madonna.

Such an extraordinary ruse certainly deserves a spectacular ending, and this is precisely what happened. Some time after Spragg became ensconced in his temple, the postman who delivered mail to the New Jerusalem community fell in love with one of the visiting virgins. In a fit of jealousy, the postman broke into the temple, fired three shots, and then fled. However, since none of Spragg's followers dared to enter the temple, it was left to the next night's virgin to report what she discovered. She finally emerged from the temple carrying a 'revelation' from the Invisible Presence which read, 'Fear not my people. Thy God is immortal.'Following this incident, some time went by until Spragg's daughter-in-law publicly accused her husband Obadiah, as well as Spragg Sr's other son Jared, of taking it in turns to sneak into the temple to have sex with the virgins. 'It took two of them to fill the place of the Prophet,' she cried, referring to Spragg Sr's prodigious sexual stamina. Four members of the community decided to settle the matter once and for all, and to enter the temple – despite vociferous remonstrations from Obediah and Jared. After an hour they came out with the news that it was totally empty. Although Obediah claimed that his father had been taken to heaven in a chariot of fire, without its charismatic leader the community folded.

In addition to the occasional extreme example, such as that of Cyrus Spragg, every human society is laden with proof that the our species has a long history of promiscuity. Even when humans have 'settled down' in conjugal bliss, their promiscuous heritage remains ever present – threatening to break through the monogamous veneer given any opportunity in the form of adulterous liaisons, affairs and secret fantasies. And it is certainly not just men who are doing this – women, too, are just as promiscuous if given the chance. One particular study in Romford, England, just after the Second World War, accidentally revealed that at least 25 per cent of children were not fathered by the legal husband, while a 1980s survey of readers of *Cosmopolitan* magazine reported that 54 per cent of all married women had participated in at least one affair. For nearly a hundred years, sex surveys have repeatedly shown that between a quarter and three-quarters of all married men and women are unfaithful at some point during marriage. The magazine *Marriage and Divorce Today* even went so far as to claim that 'seventy percent of all Americans engage in an affair sometime during their marital life'. And before this information is used as proof of Western society's miserable and dysfunctional condition, a recent survey of past and present cultures throughout the world, covering a huge variety of different marriage traditions, found examples of adultery in every single one. With recent increases in sexual freedom, humans have not reverted to a mythical polygamous mating system, with men desperate to collect and guard as many women as possible at the same time, and women wanting to join multi-female groups and live under a single powerful male. Instead, we have merely unleashed a strong desire to have sex with a variety of different partners. Like it or not, from mistresses to prostitutes, one-night stands, sex beneath desks at office Christmas parties, daydreaming and night-time fantasies, nearly all of us are, in some form or another, attracted to sexual experiences beyond monogamy.

Just about every society in the world has at some stage in its history possessed rituals that allow married couples to indulge in extramarital sex. For over a hundred years, anthropologists have recognised the importance of promiscuous festivals as safety valves to prevent the build-up of sexual tension and encourage members of a tribe to placidly accept the rules and regulations of ordinary life. This function was readily acknowledged by many of the tribes themselves, such as the Murngin tribe of Australia who had regular 'Gunabibi' ceremonies which included ceremonial extramarital intercourse. The attitude of the Murngin was that by allowing extramarital sex during in this socially regulated way, it avoided adulterous liaisons occurring at other times. In the words of a tribe member, 'It is better that everybody comes with their woman and all meet together at a Gunabibi and play with each other, and then nobody will start having sweethearts the rest of the time.' Similar promiscuous festivals have been observed all over the world, including West and East Africa, New Guinea, Hawaii, Tahiti and New Zealand. In each place, licence is given for temporary promiscuity to balance the strict control of mating at all other times.

Of all the possible outlets for our species' promiscuous urges, few remain socially acceptable today. Although they were often highly revered in past times, prostitutes have nearly everywhere been reduced to social pariahs, despite their undoubted role as pressure relievers in today's sexually confused society. There is, however, one unwitting group of people who are today doing a sterling job defending the last remaining bastion of socially regulated promiscuity. Within housing estates, condominiums and suburbs across the world, they are practising a historic and highly honourable tradition that goes back countless thousands of years. They are the swingers. Of all the ways that our species has invented for reconciling our monogamous and promiscuous proclivities, wife-swapping along with wife-lending, is perhaps the oldest and historically the most widespread of all. For the Inuit, wife-swapping used to be *de rigueur*. At times, it resembled friends swapping videos, with wives being exchanged merely for entertainment and a variety; at other times, it was more like swapping tools, as wives were sometimes loaned out for whole seasons if they had a particular skill that a friend needed. Perhaps unsurprisingly, given the male chauvinism that once pervaded anthropology, there is little mention of the control that wives had over such arrangements. All over the world, it used to be commonplace for husbands to offer their wives to guests for the night. In Siberia, for example, tsarist officials complained of being pestered by local men to sleep with their wives, and then being overwhelmed with congratulations when a son had been born. Ancient Arabian customs were much the same, and still existed in the early nineteenth century when one explorer wrote, 'Custom requires that the stranger should pass the night with the host's wife, whatever may be her age or condition. Should he render himself agreeable to the lady, he is honourably and hospitably treated; if not, the lower part of his "abba" or cloak is cut off and he is driven away in disgrace.'

From Ghanaian yam festivals and Irish sexual hospitality to Babylonian sacred prostitution and modern-day swingers, the world is bursting with examples that show just how deep and ancient is our promiscuous urge. It is an urge, incidentally, that is reflected in our anatomy – in particular by men's somewhat large testes – but more of this later. Yet, despite the strength and depth of this urge, is it really the major driving force behind our species' mating system? It is, of course, risible to suggest that humans treat each other merely like ships that pass in the night. Promiscuity might represent a continual temptation, but it has undoubtedly been pushed into second place by a more recent adult emotion that is of even greater intensity.

SHALL I COMPARE THEE
TO A SUMMER'S DAY?

On the surface, men appear to be driven by their genitals, to father as many children as possible by whatever Machiavellian means, and to invest as little energy and effort as possible into each one. Women, on the other hand, who are limited to having one child at a time, look for as much commitment as possible from their partner. We are apparently presented with a classic 'battle of the sexes', with men and women having diametrically opposite agendas and doing their utmost to manipulate the situation so that they get their own way.

There is only one problem with this scenario, and most people would instantly recognise it. It is simply not true. Men, as well as women, typically fall in love with one woman at a time. Ever since humans first started to record their emotions for posterity, men and women have fallen passionately and indomitably in love with each other. For over four thousand years, elders and storytellers in every corner of the globe have handed down tales of passionate love. The intensity of this emotion is frequently illustrated by stories in which couples would rather die than be separated. The dying passions of Romeo and Juliet, and Tristan and Isolde, are mirrored in the Middle East by the equally tragic story of Heer and Ranjha, and in India by Radha and Krishna, as well as couples in the *Mahabharata* and *Ramayana*. In one form or another, almost all the ancient societies – such as China, Egypt, India, Japan, Korea, Persia and other Arab countries – possess a rich collection of passionate love stories.

Each storyteller has his or her own way of describing the essence of this all-consuming and universal emotion. The eleventh-century Persian poet Fakhr-ud-din Gurgani, for example, recounts the moment when Ramin, brother to King Moubad, first set eyes on Vis, the King's bride-to-be:

He fell from his horse as mighty as a mountain, like a leaf that the wind rips from the tree. The brain in his head had begun to boil from the fire in his heart; heat had fled from body and sense from head … The rosy cheeks had turned the colour of saffron; his wine-coloured lips blue as the sky. The hue of life had deserted his face, the insignia of love appeared there in its stead.

Ancient philosophers approached love from a slightly different perspective. The Greek philosopher Plato provided his own version of a common theme – that of lovers being two halves of a whole. In the fifth century BC, Plato wrote in his *Symposium* how the world was once home to androgynous beings:

> These early beings consisted essentially of two people, folded together in a spherical ball. They could walk upright or walk about, turning nimbly on their four hands and four feet like tumblers. Eventually, these arrogant beings managed to enrage the gods. To punish them, Zeus cut them in half, 'like a Sorb-apple which is halved for pickling'. From then on, each desolate half was left to wander the earth, searching for its other half. If men and women ever did meet their 'better half', they were determined not to be separated ever again. They clung to one another. They did not want to do anything apart. They were willing to die rather than risk separation.

Scientists, meanwhile, strive to encapsulate the tangible threads of this most central human emotion. According to psychologist Elaine Hatfield and historian Richard Rapson, passionate love can be defined as:

> A state of intense longing for union with another. Passionate love is a complex functional whole including appraisals or appre-ciations, subjective feeling, expressions, patterned physiological processes, action tendencies, and instrumental behaviours. Reciprocated love (union with another) is associated with fulfil-ment and ecstasy. Unrequited love (separation) is associated with emptiness, anxiety, or despair.

However love is described, it does appear hard to reconcile this particular emotion with the behaviour of sexual Olympians such as Fei-ti, Atahualpa and Montezuma, and the existence of promiscuous festivals and swingers' orgies. Or is it so hard? If one looks closer at the relationships of these three ultra-polygamists, it becomes apparent that although they had sex with thousands of women, they were legally and emotionally bonded to a single woman – their wife – who they elevated above all others. Although Fei-ti had sex with thousands of women, he had a long-term relationship with only one. As for the other ten thousand women, they were merely a series of state-encouraged affairs – the sheer number of which caused even Fei-ti to complain bitterly – that were designed to spread his royal blood as widely as possible. Emotionally, if not physically, men such as Fei-ti could logically be described as 'monogamous' since their affections were restricted to one woman even though their genitals were not.

It is not just among the world's rich and influential humans that monogamy emerges through the veil of polygamy. Within all societies that allow polygamy, the majority of men are nevertheless monogamous. In the hunter-gatherer Bushmen, for example, at most 5 per cent of all marriages are actually polygamous, and the majority of these result from the fact that men are obliged to marry their brother's widow. In his study of 250 cultures, anthropologist George Peter Murdock concluded that 'nearly every known human society [is] monogamous, despite the preference for and frequency of polygamy in the overwhelming majority'.

So what words aptly describe the human mating pattern? According to the *Oxford English Dictionary*, the various alternatives are defined as follows:

Monogamy: **1** The practice or principle of marrying only once, or of not remarrying after the death of the first spouse. **2** The condition, rule, or custom of being married to only one person at a time. **3** *Zool.* The habit of animals, esp. birds, of living in pairs, or having only one mate.

Polygamy: **1** Marriage with several spouses, or more than one spouse, at once; the practice or custom by which one man has more than one wife or one woman has more than one husband at the same time. **2** *Zool.* The habit of mating with more than one, or several, of the opposite sex.

Promiscuity: Indiscriminate; making no distinctions, undiscriminating. Now especially characterized by casual or indiscriminate changes of sexual partner.

Quite clearly these definitions are woefully inadequate when it comes to describing the reality of human sexual behaviour. What label should be used to describe, for instance, a man who is married to one woman, but who regularly visits prostitutes? Or a situation in which many women share a single husband, but each of these women also has several lovers? Or, in the case of Fei-ti, a king who has many concubines but is married to a single woman, his queen?

Rather than embarking on a lengthy discourse on the merits of the various anthropological and zoological definitions of mating and marriage systems, or, as Oxford anthropologist Vernon Reynolds reluctantly admitted, being forced to 'muddle through', far preferable is to adopt an alternative and more inclusive description of human mating patterns based on emotions and biological drives.

For a start, love – of the conjugal type that Plato defined in terms of feeling helpless and lost without the other half – is perhaps the most obvious and demanding of all human emotions. Cupid's arrow flies in one direction and very rarely, if ever, in two. Love can wax and wane, and can certainly be transferred from one partner to another, but nevertheless it can only be directed towards one person at a time. In the words of the Indian poet Kabir, 'The lane of love is narrow / There is room only for one.' Emotionally, the human animal is driven to form a pair-bond with a single mate. Regardless of the number of prostitutes that are visited, the number of affairs that people have, or the number of concubines that kings have, the typical condition of adult humans is to seek the reassuring affection of one partner, companion or 'soul mate' at a time. Until it is either dissolved or transferred, this emotional partnership remains distinct from all other relationships – whether these other relationships involve sex or not.

Where then does this leave us in terms of a description of the human mating system? The answer is straightforward. Since humans form a distinct emotional bond with a single partner, we have to be described as 'monogamous' in the sense of having only one 'mate' or 'partner' at a time. With regard to so-called 'polygamous' human arrangements, the evidence strongly suggests that these are in truth founded around monogamous relationships. There are two monogamous scenarios that can easily give the illusion of polygamy. The first is that of the 'live-in lovers', in which society permits a man to satisfy his craving for extramarital sex by allowing him to marry a series of younger women. According to Mormon women such as Fanny Stenhouse, this describes the Mormon situation in which polygamy is nothing more than male licence for adultery. Within such polygamous marriages, it is common for the senior wife to choose not only her husband's younger wives, but also to draw up a schedule for his sleeping arrangements. A few years ago, I witnessed such a marriage system first-hand while visiting a famous Cameroonian pop star called Mongo Faya, who had nearly a hundred wives. The second scenario that gives the false impression of polygamy is where a man progresses from one wife to the next, in a series of monogamous relationships, but in which the 'ex-wives' remain in the household along with their children, and continue to be looked after by him. The celebrated Polish anthropologist Bronislaw Malinowski – a man who personally observed virtually every culture in the world – was firmly of the opinion that this was the primary explanation for polygamy. He insisted that, 'Monogamy is, has been, and will remain the only true type of marriage … as a rule a polygamous cohabitation is a successive monogamy'.

A vivid illustration of our species' ineluctable urge to form pair-bonds is provided by the extraordinary social experiment of John Humphrey Noyes in the 1840s. Born in Brattleboro, Vermont, in 1811, Noyes was Christian zealot, communist and creator of 'Perfectionism' – a religious and social doctrine that was dedicated to the avoidance of sin, including a rejection of personal possessions and monogamous marriage. In place of monogamy, Noyes proposed 'complex marriage'.

When the will of God is done on earth, as it is in heaven, there will be no marriage. The marriage supper of the Lamb is a feast at which every dish is free to every guest. Exclusiveness, jealousy, quarrelling, have no place there, for the same reason as that which forbids the guests at a thanksgiving-dinner to claim each his separate dish, and quarrel with the rest for his rights. In a holy community, there is no more reason why sexual intercourse should be restrained by law, than why eating and drinking should be – and there is as little occasion for shame in the one case as in the other ... The guests of the marriage supper may have each his favourite dish, each a dish of his own procuring, and that without the jealousy of exclusiveness. I call a certain woman my wife – she is yours, she is Christ's, and in him she is the bride of all saints. She is dear in the hand of a stranger, and according to my promise to her I rejoice. My claim upon her cuts directly across the marriage covenant of this world, and God knows the end.

Within his commune at Oneida, New York, Noyes banned romantic love – considering it to be selfish and shameful – and instead encouraged his followers to have sex with a variety of different partners without forming permanent relationships. In heaven, Noyes asserted, monogamous relationships cannot exist since this would isolate certain saints. If, therefore, the kingdom of God were now among men, then sexual relations must not be limited to a single mate. Of course, on earth if not in heaven, there was the problem of pregnancy. So, to avoid an excessive number of births, Noyes introduced 'male continence' – a practice that allegedly involved the men constricting their sperm ducts during intercourse, thus making it possible to have an orgasm without ejaculating. The Oneida experiment was essentially a group marriage where everybody had sexual access to everybody else, and all things were shared equally between members of the commune. All things, that is, except for sexual access to nubile virgins – which was the prerogative of the 'Central Members' who, Noyes argued, maintained a closer association with

God than did the others. It was the age-old communist paradigm of all people being equal, but some being more equal than others. So, what was the fate of this Utopian society?

The answer is that, in 1879 – that is, thirty-two years after the commune started – a faction led by a new convert, James Towner, challenged Noyes' authority by accusing him of raping young girls. Indeed, a gynaecologist later confirmed that young girls were a mere thirteen years old when Noyes and his Central Members first had sex with them – although it has to be said that Towner was almost just as guilty of having sex with such youngsters as Noyes – albeit second in line. As a result of Towner's challenge, Noyes decided to flee rather than risk possible mutiny and humiliation. But what is most interesting about this experiment was Noyes' failure in preventing his members from falling in love and forming clandestine pair-bonds. Even Noyes admitted that it was a constant struggle to keep couples apart, and on occasions he resorted to sending either the man or the woman to a distant community in Connecticut to break up the partnership. Predictably, as soon as Noyes left the commune, the remaining members changed the rules to allow monogamy, and within a very short space of time thirty-seven couples got married. To give him his due, as soon as he realised that the Oneida commune was disintegrating, and despite his aversion to monogamy, Noyes did agree to unite several couples in marriage to prepare them for the outside world. The message from the Oneida experiment is clear – even when faced with threats of expulsion, the wrath of God and intense social pressure, monogamy cannot be extinguished from the human psyche. As the anthropologist Margaret Mead put it, 'No matter how many communes anybody invents, the family always creeps back.' Humans are inescapably driven to become emotionally attached to one other individual – a process that involves the most potent and often most creative of all our species emotions – love.

Shall I compare the to a summer's day?
Thou art more lovely and more temperate:
Rough winds do shake the darling buds of May,
And summer's lease hath all too short a date:
Sometime too hot the eye of heaven shines,
And often in his gold complexion dimm'd;
And every fair from fair sometime declines,
By chance, or nature's changing course untrimm'd;
But thy eternal summer shall not fade,
Nor lose possession of that fair thou ow'st,
Nor shall Death brag thou wander'st in his shade,
When in eternal lines to time thou grow'st;
    So long as men can breathe, or eyes can see,
    So long lives this, and gives life to thee.

William Shakespeare

THE PARADOX OF LOVE

So there, with the final endorsement of William Shakespeare, we have it. Humans, despite an underlying urge to spread their sexual favours as widely as possible – and even indulge in opportunistic promiscuity – nevertheless become besotted with one other. Like it or not, through the psychological mechanism of love, they are drawn together to form emotionally bonded pairs. Of love's existence there is no doubt. But despite this, there still remains the major concern that it is not an obvious way for a mammal, such as we are, to behave.

Among mammals in general, monogamy rarely makes sense, particularly for males, this being illustrated by the fact that a mere 3 per cent of mammal species form pair-bonds. Furthermore, within the primate group of mammals, to which we belong, all the other monogamous species, such as gibbons, are strictly territorial – in other words, they live isolated in pairs and fiercely fight off any other members of their species that enter their territory or come anywhere near their mate. Under these circumstances, monogamy clearly makes sense, since every male can be virtually 100 per cent sure that his sexual partner gives birth to his offspring, and so it is in their interest to devote all their time to these young. However, this situation could hardly be more different from that of our own species, which forms societies in which many males and females coexist, and in which there can be very little certainty among males that any female's offspring is his. Uniquely among primates, we are monogamous, and yet we do not live in totally isolated pairs. And this is no new situation for members of our species. For hundreds of thousands of years, our male ancestors would have almost certainly gone off hunting leaving their females behind unguarded. In today's society too, most monogamous couples go their separate ways during the day. Yet still, despite the possibility that they are not the biological father of their wives' children, men nevertheless become strongly bonded to individual females and their offspring. It is a bizarre state of affairs.

So, how and why did pair-bonding evolve, especially if the human male has so little reassurance that he is not wasting all his time and energy looking after the children of other males? How did it come to largely replace our ancestral promiscuous mating system that seems to suit our mixed-sex social groups far better?

Most evolutionary biologists would declare in confident unison that love is an emotion that has been selected over tens of thousands of years to ensure that men and women stay together for long enough to raise their young. Well, it certainly sounds good, but is it true? To what extent is it really the case that men contribute significantly to the survival of the offspring of a single female – and does it really offset the advantages of mating with a number of different females? The old argument that the human male's contribution has been vital to ensure that his wife and offspring have sufficient food has been undermined by recent discoveries from a number of different cultures. If love is, indeed, the evolutionary mechanism that ensures that a father stays for long enough to ensure the survival of his wife's children, two predictions should be borne out. Firstly, men should hunt in such a way as to maximise the amount of meat they bring back. Secondly, they should make sure that members of their own family are well fed before offering the meat to others. But do either of these things happen?

Paraguay's northern Aché Indians are typical hunter-gatherers. The men specialise in hunting large mammals such as peccaries and deer, while the women pound starch from palm trees, gather fruits, collect insect larvae and care for their children. This much Kristen Hawkes from the University of Utah already knew before she joined the tribe to find out just what the men contributed to their family. From calculating the calorific value of the food the men brought back, Hawkes's first shock was that men actually brought fewer calories back to the village, through their hunting, than if they stayed in the village to help the women. It was not even a matter of the men bringing back essential protein since this was quite adequately provided by the fish, insect grubs and nuts that the women gathered.

Immediately, it was apparent that hunting was not just about bring-ing home the calories. However, the greatest surprise was yet to come. When Hawkes observed men coming back from the occasional hunt that yielded a large animal, she found that instead of giving the meat to their wives and children, these hunters distributed the vast majority around the rest of the village. In a similar study of the Hadza tribe in Tanzania, Hawkes discovered much the same situa-tion. Hadza men not only face enormous risks when they go out hunting in the African bush, but they do this despite failing to catch anything on ninety-seven days out of every hundred spent hunting.

Hawkes then went one step further to prove that the Hadza men hunted for reasons other than simply to provide their family with the maximum amount of meat – she asked the hunters to try catching small game such as guinea fowl using snares and traps. The result was that they came back with meat on far more days than previously. From this, Hawkes concluded that if Hadza men were interested only in the welfare of their children, then snaring was a much better way of providing meat on most days – rather than turning up with huge amounts of steak once a month. The conclusion is potentially alarming to those who see the husband's role as being the reliable provider of the Sunday roast. It suggests that something other than the best interests of his wife and children lies behind Aché and Hadza men's preference for big-game hunting.

Why, then, do Aché and Hadza men hunt if not to provide their family with meat? The answer comes from two clues. The first is provided by the Aché women themselves, who admitted that they preferred to have extramarital sex with the best hunters. The second clue is provided by our close relatives, chimpanzees, who also share the meat within their group. Scientists at Gombe National Park in

Tanzania were amazed when they first observed hunting parties of chimps racing through the trees in a highly coordinated manner, as they pursued their favourite prey of red colobus monkey. But despite many attempts at catching an adult monkey, the most common prize was merely a baby, which when shared between a large group of adult chimps provided more of a finger buffet than a full banquet. The scientists then noticed something that they found extremely curious, which was that the chimps sometimes chose to ignore colobus monkeys that were right in front of them, preferring to carry on chewing vegetation. Looking for an explanation for this ambivalence, they hit on the reason – it was simply that there was no sexually receptive female in the group. If we have seen it once, we have seen it a million times. If there is a sexually available female around, males go to virtually any length to impress her – and in the case of chimps this can mean chasing through the trees to present her with a bit of monkey. What is more, the male chimp who is most generous with his piece of dead monkey is also the one who is most likely to be allowed to have sex with the receptive female. And so the reason for hunting – at least in the hunter-gatherer societies so far studied? It seems that it has more to do with impressing members of the opposite sex than fulfilling family responsibilities.

What general conclusions though can be drawn from these findings? The evolutionary biologist Jared Diamond strongly suspects that hunters throughout the world may have agendas that are very similar to those of the Aché and Hadza men.

From my own experience in New Guinea, Hawkes's conclusions are likely to apply even more strongly there. New Guinea has few large animals, hunting yields are low, and bags are often empty. Much of the catch is consumed directly by the men while off in the jungle, and the meat of any big animal brought home is shared widely. New Guinea hunting is hard to defend economically, but it brings obvious payoffs in status to success-ful hunters.

And what about the possible relevance of Hawkes's findings to our own society? If we substitute money for meat, it is then possible to apply the same investigation to a Western urban situation. What exactly does the Western man do with his hard-earned 'bacon'? This time, Diamond hedges his bets.

> Perhaps you're already livid because you foresaw that I'd raise that question, and you're expecting me to conclude that American men aren't good for much. Of course that's not what I conclude. I acknowledge that many (most? by far the most?) American men are devoted husbands, work hard to increase their income, devote that income to their wives and kids, do much child care, and don't philander.

Personally, I see a good deal less difference between the Aché hunters and Western husbands than Jared Diamond. Far from being the norm, male fidelity is largely an illusion. If men do not actually achieve infidelity – although, as extramarital sex statistics show, vast numbers do – they are invariably fantasising about it. Then there are the vast numbers of men who walk out on their wives as well as children. In the United States, single parents outnumber co-parents, and most single parents are women. Even among devoted couples, there is a strong tendency for men to buy flashy cars, or other appurtenances of wealth, to attract sexual interest. Fundamentally, expensive items ulti-mately serve to promise money – the Western equivalent of meat – in return for sex. There are also time-budget studies that show that American working women spend on average twice as many hours on their family responsibilities as their husbands. However, when husbands are asked to estimate the amount of time that they and their wives devote to house and children, they tend to exaggerate their own contribution while underestimating their wives'. And there is little reason to believe that men's contribution is any higher in other coun-tries – in fact, American men may well spend substantially more of their resources on their household and children than most.

Monogamy still remains such a mystery. On the one hand, it is a highly unusual mating system for a primate that lives in large mixed-sex groups because of the high chances of cuckoldry. On the other hand, it now appears that despite being monogamous, men would far rather use whatever resources they have – such as meat or money, depending on the culture – to have as many extramarital affairs as they can get away with, rather than donate it to their family. This behaviour simply does not seem compatible with the traditional argu-ment that monogamy evolved in our species because of the vital contribution that males make towards the survival of 'their' offspring. In fact, it almost seems that monogamy has been imposed on males whether they like it – or is in their best interest – or not.

Evolution, of course, is far from simply being about what is in one sex's best interests. Any man, for instance, who thinks that he is acting solely in his own best interests is deluding himself. Instead, men oper-ate within the confines of what women – be it their sexual partner or other potential sexual liaisons – want them to do. Females have just as much, if not more, part to play in deciding male behaviour as males themselves – as already shown in baboons and chimpanzees where females have considerable influence in deciding which males get to mate with them. Regardless of what males 'want' to do, females will attempt – by judicious mate-choice – to select males that contribute as much as possible to them and their offspring. From this perspective, men becoming besotted and emotionally bonded to their sexual partner starts to make sense. However little men contribute towards the feeding and protection of a female's young, it is better than nothing. But if females were indeed behind the evolution of the pair-bonding emotion we call 'love', the sixty-four-thousand-dollar question is – how could they possibly have achieved it?

chapter five

the evolution of love

With three children in tow, Anne steered a pushchair carrying her fourth child towards her front door. As she leant forward to put the key in the latch, she noticed a slip of paper that had been attached to the doorbell. It was a note from Stephen, one of her neighbours, that said, 'See you tonight, OK?' With the three children becoming impatient, Anne opened the door and bundled everyone inside. She hastily settled the children in front of the television, and headed towards the kitchen to put away the shopping. However much she tried to concentrate on household matters, her mind kept on returning to the possible consequences of the night that lay ahead. Stephen was just one of the men who visited her regularly, but although he was very kind and provided her with plenty of food, she was worried about what would happen if the whole village got to hear about her liaison. The main problem would be Alan, since he had already let it be known that he wanted to be first in line when she was next ready to have sex. Although Alan and Stephen were not directly in competition, there had been a certain amount of tension between them ever since Alan killed one of Stephen's close friends during a dispute over village leadership. As Anne looked out of the kitchen window, she realised what lay behind the timing of Stephen's note. Further down the street, there was a gathering of men around the house of a young girl who had just arrived from a neighbouring village. Clearly, this newcomer was prepared to have sex, and Alan was suitably preoccupied as he staked his claim to be at the front of the queue.

Despite the extraordinary power of human imagination, it is surprisingly difficult to picture a society in which people behave according to fundamentally different rules from those that we are used to. Yet there is one particular imaginary scenario – in which the human species is exclusively promiscuous – that is not so inaccessible to our minds, largely because of its ubiquitous existence beneath the respectable monogamous fabric that we see around us every day. In a promiscuous human world, all women would be 'prostitutes' – they would survive, and raise their children, by exchanging sex for various favours, such as money or food. Men, on the other hand, would visit as many women as their status or wealth would allow, without forming an emotional relationship with any of them. Once a man had received sex, he would not return to a wife or girlfriend back at home, but instead would live among other males, working to accumulate wealth, improve his status and defend the village along with his mating rights within it.

But there is one fatal flaw to this promiscuous scenario, and it is as simple as it is devastating. Without the evolution of monogamy, there would be no humans in the first place. Far from being an optional accessory, monogamy – or more specifically 'love' – has been central to the emergence of our unique species. Moreover, if it had not been for a significant vagary in the African climate over 2 million years ago, neither love, nor humans would have evolved.

THE DAWN OF EQUALITY

Around 2.5 million years ago, mighty Africa was plunged into yet another Armageddon period. It was the second of the two major drying-up periods that Africa has experienced over the past ten million years – the first having occurred some four million years earlier. This time the drought was to last about 0.5 million years. Immense climatic fluctuations gripped the world as ice caps advanced and retreated every few thousand years, bringing about calamitous cycles of cooling. Many of the African forests that managed to survive the first period of drought now succumbed – to be replaced by dry savannah – while millions of animals perished. But just like the first period of drought, which gave rise to the first highly social upright apes, this second period was to have an equally profound effect on our ancestors. By the time this latest environmental nightmare was over, at just under two million years ago, our ancestors not only possessed brains that were far larger than ever before, they had also spread right across the globe, to Europe, China and even Java.

What could possibly have happened in this traumatic African episode that caused the genesis of this new and highly successful ancestor? And why was this ancestor's brain so much larger than before? The answer to this riddle lies hidden in the bones. What is immediately apparent from the fossil records is that these new ancestors had lighter and more delicate bodies than before, and they were more agile and long-legged than any of their predecessors. Describing the body of the old type of ancestor, one anthropologist noted that it 'would not have been able to lift its thorax for the kind of deep breathing that we do when we run. The abdomen was potbellied, and there was no waist, so that would have restricted the flexibility that's essential to human running.' Whereas the old type was heavy-bodied and designed more for sitting, the new type was an athlete and runner. Supporting this view is the finding that the new type also had balance organs in its inner ear that are identical to ours, rather than the chimpanzee-like structures of the old ancestor, giving them much better balance when moving fast on two legs. Richard Leakey, who discovered a near complete skeleton of a six-foot-tall boy belonging to this new type, has even gone so far as to describe how, 'Suitably clothed and with a cap to obscure his low forehead and beetle brow [heavy protruding eyebrows], he would probably go unnoticed in a crowd today.' These new ancestors' heads were also very different. They had flatter faces, smaller teeth, shorter jaws and more rounded skulls. In fact, apart from their smaller brains, these new ancestors were essentially the same size and shape as we are today.

The catalogue of changes that occurred in our ancestors two million years ago were astonishing in one particular way: they were all linked to persistent or exaggerated infantile features. The process that started four million years earlier – but which had been on hold since then – had started again, and to massive effect. The new ancestor was essentially an even more infantised version of the old one. Speaking of this new type of ancestor, the anthropologist Colin Groves described, in anatomical language, the changes that occurred:

Almost all its uniquely derived character states are neotenous [infantile]. These are the simplified morphology of P_3 and the canines; reduction of upper canine size; the gracile cranial form, without a defined supramastoid crest; and shortened basicranium, with basion well in front of bitympanic line.

Whereas the first round of infantising created an ape that stood up on its back legs, the second round created one with a more delicate body, a bigger brain and smaller teeth – one that now bore an even closer resemblance to an infant ape. But why was the infantising process restarted now? What, for that matter, were the advantages of shedding even more of our ancestors' mature characteristics?

As for what kick-started the infantising process back into action – of this, there can be little doubt. Much like the first round of infantising, the second round also occurred during a stressful environmental change. Once again, the environment was becoming increasingly drier, and this would have had the unavoidable effect of making food even more scarce than it was before. Whereas before the drought there was sufficient food available for females to feed themselves and their offspring, albeit with difficulty at times, now the situation was desperate. To make matters even worse, with ever decreasing amounts of food, the competition would have become intense, both within each species as well as between different species. The result was that, among the various upright apes in Africa at the time – of which there were at least fifteen species – there was an enormous selection pressure for a population in which males contributed towards the feeding and protection of the offspring through the provision of food and protection. Until now, being of promiscuous origin, the males would have mated and then left the females to look after their own young. To make matters worse, there was still the problem of paternity certainty, with males having little assurance that any of the offspring were actually theirs.

So what happened next? Did the males, as might be expected of a species that lived in a large mixed-sex group, simply share more food with whichever female they copulated with? Or did they opt for the high-investment strategy of feeding the young of one particular female, despite having no idea of whether these young were fathered by him? Perhaps surprisingly, the evidence suggests that it was the latter.

Just occasionally, one encounters a line of reasoning that is so obvious that we kick ourselves for not seeing it before. At once, we are dismayed by our lack of perspicacity and delighted by a new perspective. Such may well be the case for the argument that counters the claim that, since the dawn of human time, males went out hunting or scavenging, while females collected plants and insects. The revealing question is simply: what had to happen before our male ancestors could possibly spend so much time roaming rather than eating? The answer is that, for males to be able to devote vast amounts of time to searching for meat – be it hunting or scavenging – our female ancestors must first have started sharing their food with them. Hunting or scavenging is simply far too unreliable and risky a way of finding food to survive on alone, and certainly not to be undertaken if there is no back-up source of food waiting at home. Even in modern hunter-gatherer societies, with their sophisticated spears and slings, men rely on their wives to provide food on most days, since the vast majority of hunting trips are unsuccessful. As mentioned earlier, a study of the hunter-gatherer Hadza tribe of Tanzania found that men returned empty-handed on ninety-seven out of every hundred days spent hunting. Similarly, in the Kung tribe of South Africa, women's plant-gathering activities provide up to 80 per cent of the diet. The moment females start sharing their food with males, the whole nature of society changes.

So how is this logic relevant to the emergence of our new ancestors around two million years ago? The reason is because the transformation of the old-style ancestor to the new athletic ancestor coincides with the arrival of stone tools that were capable of paring meat and tendons away from bones, as well as extracting the marrow. According to the previous argument, for serious scavenging to be possible, division of labour must also have emerged. By being provided with food by the females, the males – and it would almost certainly not have been females considering the perpetual burden of pregnancy or child-carrying for adult females – were able to spend a considerable amount of time searching for lion or leopard kills to scavenge. This certainly does not mean, however, that males soon put their new-found freedom to good use by forming sophisticated hunting parties while women stayed at home. There is no evidence of hunting, using projectile weapons, until a mere 50,000–100,000 years ago – that is, nearly two million years after the period in question. These ancestors were primarily scavengers. Indeed, further support for a division of labour between the sexes comes from the discovery that they carried their stone tools for large distances – something that would have been very difficult if the whole group, women and children included, moved en masse.

The emergence of food sharing about two million years ago is supported by anatomical as well as behavioural changes in the new ancestors. For the first time in human evolution, the males and females were roughly the same size. Rather than the males getting smaller, it was the case that our new female ancestors were 60 per cent larger than before. In other words, whatever had previously limited the size of the females – perhaps a need to breed early combined with the huge energy drain of raising young alone – had been lifted. With a 'father' present, the chances of survival of both mother and baby would be considerably higher, with the result that the female strategy could change slightly. Now, it was better for females to start reproducing later in life, thus growing larger, and devote a greater amount of their time to a smaller number of babies. With the advent of the nuclear family, came far more equality between the sexes – including a larger female.

But still there exists the puzzle of why males exchanged an independent way of life for one that involved sharing food with a female. In fact, regarding the major problem of paternity certainty, the situation had become even worse now that some males were going away on long scavenging trips. Mate-guarding was therefore out of the question as a means of ensuring exclusive sexual access to a particular female. At least before this the males and females would have grazed alongside each other and thus had a rough idea what the others were up to. So how did evolution, or perhaps females – because it is arguably they who gained the most from the change – transform a fundamentally promiscuous species into one that is apparently monogamous?

BABY LOVE

Once again, the most productive method of uncovering the events of two million years ago is not only to look backwards, but also inwards and sideways. While the bones of our ancestors, and the remnants of the environment that they lived in, can tell us a certain amount about the way they would have lived, our own present-day bodies, as well as those of our close relatives, also provide us with a historical record of our past. Regardless of the present circumstances of an animal, its body and behaviour inescapably contain elements that exist through simple force of history. The puzzle is to correctly link specific features to specific dates when they first emerged.

The first clues have already been mentioned – the bones. The fact is that the bones of the new ancestors had been radically infantised. These ancestors possessed an array of exaggerated baby features. While some characteristics were simply retained from infancy – such as the relatively large rounded head and the flat face – others were more the result of highly extended infantile growth patterns – for example, the long legs. But as species as diverse as bonobos and dogs illustrate, infantile bodies and behaviours often go hand in hand. If this was indeed the case in our ancestors of two million years ago, are there any infantile behaviours that, if extended into adulthood, might explain the emergence of pair-bonding and monogamy in an otherwise promiscuous species?

Of all the relationships that exist in nature, none is as powerful as that between a baby and its mother. Infantile attachment reigns supreme among the animal kingdom's most powerful bonds. From the moment that a mammal is born, it is essential that it develops an obsessive attachment to its mother's voice, as well as her scent, body and face. To an infant, the mother is its main and possibly its only source of pleasure, protection and comfort. (Of course, a carer other than the mother can fulfil the same role if he or she has looked after the infant from a very early age.) Biologist Leonard Rosenblum has pointed out that it is not surprising that primate mothers and infants are so tightly bonded, not only because separation in the wild is generally lethal, but because infants are in continual danger from poisonous plants, insects, reptiles and predators. To ensure survival, therefore, an intense 'desire for union' is a vital feature of young primates. Likewise, from the moment that her baby is born, it is essential that a mother too develops an obsessive attachment to her baby's sound, scent, body and face. Only when her baby is close to her and safe within her arms, can a mother be fully relaxed and comfortable. For both mother and child, there is an indefatigable urge to attain close proximity with the other. In platonic terms, it is as though they are two halves of a whole that yearn for reunion – two moieties that are destined for eternal misery so long as they are apart.

Although the traditional explanation was that this infantile attachment was based entirely on the provision of food, the classic experiments of Harry and Margaret Harlow in the 1970s using rhesus monkeys showed that this was far from the truth. The Harlows' experiment was as shocking-looking as it was impressive. They first separated monkey infants from their mothers, and then placed each of these infants in a cage containing two types of artificial 'mothers'. One 'mother' was made from cold wire mesh with a milk bottle inserted into it. The other was warm and towelling-covered, but lacked a milk bottle. If the early researchers were right, then the infants should become attached to the wire 'mother' that contained the milk bottle. But they did not. While they were willing to suck milk from the wire mother, they spent nearly all their time clinging to the warm mother that lacked milk. To confirm their findings, the Harlows then introduced a mechanical toy bear into the cage, which banged loudly on a drum. Predictably, as soon as the bear made a noise, the infants rushed up to their cloth 'mother'. But what if the cloth mother was removed, leaving only the wire mother that contained the milk bottle? This time, when the bear started banging on the drum, the infants simply cried and ran frightened around the cage. They totally ignored the wire mother. The conclusion is therefore that, yes, primate babies need food, but primarily they crave warmth and comfort.

The relevance to the situation two million years ago is plain for all to see. Our ancestors' infantile anatomy – which is itself a strong indicator of infantile behaviour – coupled with the emergence of adult pair-bonding, strongly indicates that infantile dependency is at the heart of adult love and monogamy. Not only did our new ancestors extend the physical aspects of infancy into adulthood, they also extended the emotional ones too. By failing to develop beyond the stage of being dependent on a carer, adult love and dependency on a single other individual was born.

Indeed, psychologists and psychoanalysts have long recognised the link between an infant's craving for its mother, and an adult's love of their partner. The psychoanalyst John Bowlby remarked that 'in terms of subjective experience, the formation of a bond is described as falling in love, [and] maintaining a bond as loving someone.' Clearly, infant dependency and adult love share the critical features of an intense desire not to be separated from one specific individual and a longing for attention. The Japanese psychiatrist L. Takeo Doi argues that one key to understanding Japanese culture is to understand the concept of *amae* – a word that is derived from a verb that means 'to depend and presume upon another's love'. Early in life, *amae* dominates a child's relationships, causing it to cling to its mother, be utterly dependent on her and resist separation at all costs. In adulthood, Doi considers that '*amae*, generally speaking, is an inseparable concomitant of love' – when people's longing for *amae* is thwarted, they often behave like resentful children by either sulking, throwing a tantrum, or feigning indifference. According to Doi, the Japanese assume that *amae* will continue to cement their most intimate relationships throughout their lives.

Going one step further, psychologists have also suggested that children's early attachment experiences can have a strong influence on the type of loving relationships that they have later in life. Adult love, in other words, being a continuation of the same drive that binds infants to mothers, carries on where mother–child relationship ended. Elaine Hatfield and Richard Rapson provide their own version of such a phenomenon:

> Children's early patterns of attachment should influence their adult attachments. For example, we have observed that children are likely to become *securely attached* to their mothers if they are allowed to be both affectionate *and* independent ... such children should mature into secure adults who are comfortable with intimacy and are able to trust and depend on those they care for. Children may become *anxious/ ambivalent* if they have learned to be clingy and dependent, or fearful of being smothered and restrained, or both. Such children should become anxious/ ambivalent adults who fall in love easily, who seek extreme levels of closeness and are terrified that they will be abandoned.

The infantile origin of adult romantic love is tenderly illustrated by courting couples all over the world – as they re-enact virtually every behaviour that occurs between mother and child. They cuddle, like infant with mother. They kiss, like babies being fed directly from their mother's mouth. They coo and talk to each other in baby language. They call each other by childish names. They gently caress each other, put food into each other's mouths, and stroke each other's hair. Each wants to be cared for and nurtured by the other, and each is keen to indulge the other's desires.

Of all the mother–infant behaviours that human couples indulge in, kissing is perhaps the most publicly seen and clear-cut in its origin. Gorillas, chimpanzees and orang-utans – our three closest relatives – also kiss, or rather 'kiss-feed' since lip-to-lip contact is almost entirely limited to mothers who are weaning their infants off milk and on to solid food. In the early stages of weaning, the mother chews the food until it resembles a soup before putting her lips to those of her infant and then squirting it into its mouth. As time goes by, the food is chewed less and less until it contains large lumps, thereby preparing the infant for solid food that it can put directly into its own mouth. There are still some human societies today, in Africa, South America, the Philippines and New Guinea, which use this same method to wean their children from the breast. It is also the case that in some parts of the world distressed infants are sometimes calmed by an older brother or sister pressing their lips to the baby's mouth, inserting their tongue and passing a little saliva across – a comforting technique that is not so dissimilar from that of putting a nipple-mimicking dummy into a child's mouth.

Functional kiss-feeding has even been observed in adults in some parts of the world, for example, lovers from the Viru tribe in the highlands of Papua New Guinea have a tradition called 'Yangu peku' during their courtship, which involves the transfer of food from the mouth of one lover to the other. Similarly *The Kama Sutra*, a Hindu love manual dating back perhaps to the second century, also describes wine as being passed from mouth to mouth during lovemaking. In every way, our own species' sexual kissing resembles primate kiss-feeding – one partner initiating the sequence by putting his or her lips to their partner's, then pushing their tongue between their partner's lips, which in turn causes the partner to begin sucking as would an infant. The 'French kiss', in which one person fully inserts their tongue into the other's mouth, is the most extreme form of this type of infantile mimicry. Not only does it contain the element of mother–infant lip contact, it also mimics the actual transfer of material from the mother's mouth to that of the infant, with the tongue providing the stimulus of food.

Evidence that kissing is an ancient sexual behaviour of our species comes from the fact that it is far from just a Western characteristic, but rather one that is an important part of sexual behaviour in all cultures. It has been found, for example, in remote South American Yanomami tribes that have never had any previous contact with other cultures, and is depicted on Peruvian ceramics that date back to times long before Spanish colonisation. Although there are suggestions that the Japanese learned kissing from Europeans, an old Japanese quote shows this not to be the case, as it warns lovers against pressing their tongue between the lips of their partner during lovemaking, since there had been cases of women biting off the tips of their lovers' tongues during orgasm. Kissing, it seems, is yet one more example of how humans require their sexual partners to provide them not only with sexual gratification, but also with comfort stimuli that they last received from their mother. It would be fascinating to know if men and women's behaviour would in any way be affected if every time they gave their partner a long lingering kiss, they realised that they were re-enacting a widespread primate behaviour that serves to transfer food from mother to infant.

THE SIREN'S WEB

[On the subject of Odysseus' journey back to Ithaca] a favouring wind ... soon brought them to the isle of the Sirens, those
sisters of enticing song. So sweetly they sang that all who heard
them were drawn on shore to where they sat in a field of
flowers, blooming among the bones of men thus lured to their
death. But on Circe's counsel, before they came within earshot,
Odysseus stopped the ears of his men with wax, and made them
bind himself fast to the mast, charging them by no means to
unloose him, however he might beg or command when his ears
were filled with the fatal voices.

Thus prepared, winged by their oars they flew past the beach
on which could be seen the Siren Sisters, and over the waters
came their tempting strains, heard by the captain alone ... Their
song so thrilled his heart that Odysseus struggled hard to get
loose, and by cries and signs would have bidden his men undo
the cords; but they tied him up all the tighter, and deaf to him
as to the Siren music, rowed their best till they were far out of
hearing. Then only they unbound him, and took the wax from
their ears; and for once the Sirens had sung in vain.

It is time to face the vexatious issue of the evolutionary mechanism
by which our ancestors became infantised. Was it pure natural selection – in which infantile individuals were simply more successful in
raising offspring males – without males or females specifically choosing certain types of partner? Or was sexual selection involved? Was
one sex specifically targeting infantile members of the opposite sex –
like sirens ensnaring their hapless male prey – and insisting on mating

only with this type even if it proved disadvantageous to the 'victim' sex in question? Did one sex drive our species' evolution towards a condition of perpetual infancy?

To avoid mawkish sentimentality, it is as well to remember at this point that monogamy is an aberrant mating system for a primate that lives in mixed-sex groups. Indeed, humans not only live in mixed-sex groups, and are monogamous, they also tend to form single-sex cliques within these groups for much of the time. From hunting and drinking parties, to cooking and gossip groups, men and women throughout the world spend much of their time apart. In the vast majority of traditional societies, males are responsible for most if not all the scavenging and hunting, as well as protecting their group from attack by predators or neighbouring groups, while females are constrained by the demands of pregnancy and child-rearing. Far from being socially constructed, or sexist, such a division of labour is merely the result of women, not men, becoming pregnant. But for reasons given earlier, this basic social organisation does pose an immense problem for males in terms of paternity certainty. Since males spend much of their time away from females, they have little chance of monitoring any one female's sexual behaviour. Needless to say, natural selection would very quickly expunge the genes of males that gratuitously opted to spend their lives looking after another male's young, rather than his own.

But there is one evolutionary mechanism that is perfectly capable of explaining the emergence of behaviours that, at first sight, seem to disadvantage one sex or the other. It is the same mechanism that has led to the evolution of the peacock's tail – a gaudy and cumbersome structure which quite clearly increases the chances that its owner will be noticed, caught and eaten by a predator. It is sexual selection – a process that indulges the desires, however capricious, of one sex or the other. If peahens choose to mate only with peacocks with gaudy tails, then that is what peacocks must have if they are to mate. Similarly, if our ancestral females chose to mate only with males who formed close relationships with a female, then that feature will spread throughout the population in very little time. Given the puzzle of monogamy in our ancestors, it therefore makes sense to explore the possibility that it emerged not as a result of natural selection, but rather as a devious female strategy to ensure that the males helped with childcare.

If conjugal love is indeed the result of female sexual selection, it would certainly not be the only female device to limit the promiscuous tendencies of males. Unlike our closest ancestors, the chimpanzees and bonobos, human females are bizarre in concealing the timing of ovulation. Women, as if it is not obvious, do not develop large brightly coloured swellings around their genital area just as they are about to ovulate. As far as males are concerned – without the aid of thermometers, or charts that plot the progress of the menstrual cycle – sex is something of a guessing game if the objective is to make a woman pregnant. The reason for this is obvious. Just imagine the response of men in a village if it was plain for all to see exactly when each woman was at her most fertile. The effect would be to encourage all the men to compete for one woman after another – in a merry-go-round promiscuous orgy – as each approached the point of ovulation. This, however, is not the only way that human females confuse men. Instead of being willing to have sex only when they are ovulating – again, like most other primates – they invite sex throughout most, if not all, their menstrual cycle. With visual and behavioural signals of ovulation both gone, men are denied two major cues for opportunistic sex. Finally, and just to complete the 'anti-promiscuity' package, women who live in close proximity also develop synchronised menstrual cycles. On the slim off chance that men can detect when a female is ovulating, they all ovulate together, thus making it virtually impossible for men to mate with more than one ovulating female – preferably their sexual partner – within the group.

Turning to the more practical issue of attracting the more infantile members of the male population, what type of female would be most successful in attracting a male that retained a high degree of infantile dependency? Sex is, after all, a two-way process, in which males as well as females exercise a certain amount of choice. The answer is clearly that such a female must provide as many visual and behavioural cues of motherhood as possible – for example, by being a source of care, food and comfort. They must, in other words, possess a strong inclination to treat adult males as though they were their own young. To a male who develops a sex drive but nevertheless retains many of the emotional characteristics of immaturity, the 'perfect' female partner should not only satisfy his sexual desires but also his extended need for a maternal figure in his life. In the words of French psychiatrist Dr F. Vidal, 'The most passionate lover never ceases to look upon her beloved as a mother on her child.'

For evidence that infantising and female dependency go hand in hand, we need only to briefly remind ourselves of the behaviour of the highly sociable and infantile bonobo. Of all the aspects of bonobo behaviour, one of the most extraordinary is the total reversal in the status of males and females, compared with the chimpanzee. Along with the bonobo's infantile behaviour and anatomy has come the radical emergence of 'mother power'. Whereas every male chimpanzee can brutally persecute every female in his troupe with a certain degree of impunity, male bonobos would never get away with such aggression since they would be severely attacked by the females who remain dominant to them throughout life. As mentioned earlier, even when fully adult, male bonobos frequently continue to follow their mothers around the forest much as they did when they were infants. Although bonobos have been infantised to a lesser extent than ourselves, they provide the perfect illustration of how remaining infantile both reduces aggression within a group and causes males to form considerably stronger relationships with females for the whole of their lives.

BORN TO LOVE

Sigmund Freud, for all the work that he did to reveal how our infantile experiences mould our adult lives, made one significant mistake. Although he went to great lengths to explain how both sexes have lifelong infantile demands, he failed to point out that women combine this with an immensely powerful urge to treat men as pseudo-children. As women all over the world can testify, men invariably want to be cared for by their sexual partner, while women act as a mother figure towards their partner. And the reason for this attribute of the human female is simple. At the same time that human females selectively mated with infantile males, these males, in turn, selectively mated with females who offered them the greatest maternal care. Hence, as our species became more infantile, adult females became more motherly towards adult males. In fact, the effect of selection for motherly females has been to create the most extreme maternal behaviour to be observed in any species. Human females are literally the super-mothers of the animal kingdom. Long before they turn their attention towards anything as 'normal' as children of their own, they have already been applying their maternal skills to just about anything that moves, looks like it might move, does not move but somehow looks cute, or simply appears lonely. Mothering pervades every part of a woman's life from cradle to grave, and infiltrates every aspect of her day-to-day behaviour. Girls the world over show amazingly consistent nurturing behaviour, while boys on the whole show very little. The games that girls play are indicative of this powerful maternal drive, including those that involve role-playing mothers looking after babies, or nurses providing medicines and bandages to the injured. The author Thomas H. Middleton came up with an extreme scenario to illustrate how boys and girls use play materials differently:

If some mad sociologist should ever settle a thousand little boys in a compound and give them dolls to play with and give footballs to a thousand girls in another compound, I feel certain that within a few days a small minority of the girls would be kicking and throwing the footballs around, while the majority would be cuddling their footballs and scolding them for being naughty. And I'd bet … that 60% of the boys would have dismembered their dolls to use the limbs and torsos for batting the heads about the compound; and the 10% who went in for cuddling would have their dolls stolen for dismemberment by the majority.

Later in life, the human female continues to direct her maternal instincts far wider than simply towards her own children, to include orphaned or injured animals, adopted children and, most pertinently, their husbands. My eight-year-old daughter illustrates this female tendency to distribute the maternal attention widely to perfection. To her, life is a long procession of needy causes and lost souls – as she considers herself perpetually surrounded by animal, vegetable and mineral refugees in desperate need of her tender loving care and attention. Nothing is beyond nurturing, and everything requires her maternal touch. While her ten-year-old brother constructs pseudo-weapons of mass destruction, my daughter gleefully dedicates herself to looking after all that surrounds her, from polystyrene-stuffed toy animals, to hamsters, hedgehogs, woodlice, dogs, and even a parrot made from yoghurt pots and coloured bits of paper. All require feeding, cleaning and putting to bed at just the right time, so that by the end of each day, the house is littered with the debris of this inexhaustible maternal drive. My son, meanwhile, has just finished his latest diabolic construction.

There is one particularly odd behaviour, at least in Western society, which vividly illustrates women's strong desire to nurture their adult sexual partners. It is one that Lorenz Hart captured in the lyrics to his and Richard Rogers' song 'You took advantage of me'. In this song, the lyrics describe how a girl with 'the heart of a mother' passionately craves someone whom she can love. In the absence of an available man, and since the object of her affection cannot be a sister or brother, she directs her affections to her horse. The song maintains, however, that because horses are 'frequently silly', the girl goes off in search of a man to be the recipient of her love. Once a suitable man has been found, the girl – again according to the lyrics – has no choice but to love and look after him since it is in her nature to do so, hence the title of the song. Indeed, it is occasionally claimed that, as symbols of adult masculinity, horses outdo men completely. In DH

Lawrence's short novel *St Mawr*, a wife insists 'You may have my husband, but not my horse. My husband won't need emasculating, and my horse I won't have you meddle with. I'll preserve one last male thing in the museum of this world, if I can.'

Researchers have found that as children get older they increasingly turn their attention away from large animals, such as giraffes and elephants, and towards smaller, furrier animals such as puppies and bushbabies. Desmond Morris has suggested that this is because young children prefer large parent substitutes, while older children prefer small child substitutes. But there is a recurring exception to this general rule, and it concerns girls' compulsive attraction to horses. What is quite extraordinary is that as girls approach puberty, they frequently become infatuated with horses. To boys, horses are essentially the animal equivalent of racing cars – they go fast, but so do mountain bikes. To girls, on the other hand, horses can provoke intense emotional feelings and frequently become recipients of obsessive love. But why? While it is easy to understand how girls' maternal instinct will draw them towards anything resembling a small defenceless child, such as bushbabies or pandas with their large round eyes and flat face, their attraction to large imposing horses seems to defy logic. The reason, most likely, is far more to do with girls possessing a strong drive to look after adult males. Since Western society vehemently forbids relationships between pubertal girls and mature males, the best substitute for many girls comes in the form of a large horse to nurture instead. What is, of course, highly significant about this phenomenon, from the point of the evolution of human sexual relationships, is that just at a time when one would expect girls to be drawn towards babies or baby substitutes to care for, these girls instead direct their maternal instincts towards mature males or their substitutes.

THE PSEUDO-MOTHER

Women throughout the world spend vast amounts of energy administering to both their husbands and children's needs. It seems that men's proclivity for women who combine maternal care with sexual services has been highly successful. However, there is one particular aspect of human female biology that conspicuously bears testimony to thousands of generations of men having chosen sexual partners who satisfy their craving for a mother figure. I am referring to a feature that anthropologist Helen Fisher described thus:

> But what a bad design. These protuberances around the mammary glands seem poorly placed. They bobble painfully when a woman runs. They flop forward to block vision when she leans over to collect food. And they can suffocate a suckling child ... for whatever genetically adaptive reasons ... ancestral males liked females with these sensitive, pillow appendages and bred more often with sexually responsive, big-busted women – selecting for this décor.

According to one's point of view, women are either blessed or burdened with a feature that is totally unique among primates – permanent breasts. As soon as the human female reaches puberty her breasts become padded with fat, thus giving her the appearance of producing milk. Moreover, as befits such extraordinary structures, a number of theories have been proposed to explain their evolution. Some scientists have seen them as advertisements of health and well-being – as highly visible places to store fat where they will achieve maximum impact on males who are keen to mate with a female who has plenty of energy to look after his young. Alternatively, ethologist Desmond Morris views breasts as buttock mimics. According to this argument, when our species stood upright, females incidentally lost one of their most powerful sexual stimuli – the big rounded buttocks that protrude backwards when standing on all fours. Being sorely missed, males reinvented these by choosing to mate with females who had similar rounded structures on their chest. Considering the ease with which men can discern a woman's nutritional state from her overall appearance, breasts as further advertisements seem utterly superfluous. As for the buttock-mimic explanation, there are a number of objections. Firstly, without the aid of a Wonderbra to create a cleavage, breasts rarely achieve any similarity to buttocks. Secondly, women's rounded hips and buttocks are in themselves extremely effective sexual stimuli. And lastly, it is unlikely that the common ancestor of humans and chimpanzees had swellings around their rear in any case, since this seems to be a recent invention on the part of our close relatives. But perhaps more to the point, both the health-advertisement and the buttock-mimic explanations would have William Occam turning in his grave. Such explanations are unnecessary, overcomplicated and irrelevant, given the glaringly obvious fact that permanent breasts purely and simply make human females look like they are producing milk from the moment that they become sexually mature. Surely the only sensible question to ask is, 'Why are men attracted to women who appear to be producing milk?'

There is one only one credible answer to the question of why males find breasts attractive. Breasts have evolved to strengthen the bond between male and female sexual partners, by reinforcing the human female's identity as a mother figure. What the Wonderbra is actually doing is improving on nature by making breasts appear maximally tight and swollen, as though bursting with milk, and hence more maternal and even more attractive to men. In the words of T.S. Eliot, 'Uncorseted, her friendly bust gives promise of pneumatic bliss.' Given men's universal attraction to virginity, one might expect that this would weigh heavily against any feature that suggests that a female already has the burden of a child, as indicated by swollen breasts. However, so strong is men's infantile nature that even this is outweighed by their craving for a return to the comfort of the breast. It is the existence of these two rather opposing forces of attraction – virginity on the one hand and nurturing mother on the other – that explains the biologically unique female shape that is so often preferred among Western men – of a delicate infantile body coupled with enormous and apparently milk-laden breasts. Of course, as with so many other aspects of human anatomy, the original appeal of the breasts have been absorbed into the unconscious. To men now, they are simply 'sexy', along with such features as a curvaceous waist that originally indicated ease of childbearing, shiny hair that suggested a lack of parasites, smooth skin that signalled youth, as well as white teeth, clear whites to the eyes and facial symmetry that indicated a resistance to disease.

Far from being a peculiar Western infatuation, full breasts are central to sexual attraction and lovemaking in cultures across the globe. Moreover, ancient poetry and art provides compelling evidence that males have been attracted to female breasts since time immemorial. Almost as soon as sculptured art appears in the archaeological record, voluptuous 'Venus' figurines start to appear, among which is the famous 'Venus of Willendorf' from around 30,000 years ago with her hugely exaggerated breasts. Likewise, Indian literature and art in particular makes frequent reference to the female breast, including numerous mentions of how the perfect female body possesses heavy breasts, slender waist and broad hips. The Indian poet Vatsyayana, author of *The Kama Sutra*, described a variety of different ways of kissing the breast during foreplay, such as the 'Balanced Kiss' (kisses on the curve between the breast), the 'Forcible Kiss' (kisses directly on to the breast), the 'Chaste Kiss' (kisses between the breasts and the waist), and the 'Boar's Bite' – a kiss that one authority politely pointed out should be strictly avoided 'when dealing with virgins or other people's wives' since it could result in a series of bite marks being left across the breasts.

Far from venturing into the realms of Freudian psychoanalysis, this explanation for the evolution of permanent female breasts in our species is simply based on biological sexual selection. If attracting particularly infantile males was becoming increasingly important to females, in terms of maximising the number of healthy children they could have in their lifetime, then the females that were best at attracting this type of male would have passed the most genes on to the next generation. Put another way, while women selected men who behaved most like their children, these men selected women who most resembled their mothers – complete with swollen breasts. What is more, sexual selection also explains why, where nutritional conditions permit, the size of breasts can increase out of all proportion. An identical phenomenon has been responsible for exaggerated sexual characteristics in a wide variety of other species – from the dazzling and extremely costly plumage of birds of paradise, to the absurdly massive antlers of the recently extinct Irish elks. It is a process that is known as 'runaway selection'. If a particular feature such as bright plumage or permanent breasts becomes attractive to one sex, then it can easily happen that super-exaggerated versions of this feature are considered super-attractive. Huge breasts can therefore be explained as 'super-maternal' sexual signals. Even if the selected feature becomes so extreme as to be virtually dysfunctional, there is still immense selection for constant increases in size because of the success of individuals possessing them in attracting mates.

Runaway sexual selection also explains the appearance of another unique feature of our species, the large and often highly pigmented area around the nipple – the 'areola'. In other primates, the nipple merely sits at the end of the breast without any additional surrounding pigmentation. But of course, the breasts of other primates are not the subject of sexual selection. In such a situation, as occurs in our species, anything that exaggerates the appearance of these structures will be selected for – including large size and a prominent nipple. While it has not been possible to enlarge the nipple itself to provide a 'super-maternal' stimulus, since this would make it impossible for a young baby to feed, what has been possible is to make the nipple appear to be far larger by making the skin surrounding it extremely prominent. Instead of displaying a relatively small sexual signal – in the form of a small nipple at the end of the breast – the human breast gives the appearance of having a massive nipple by means of the areola. As with the enlargement of breasts, in a situation where nipples become attractive to men, then anything that makes these structures appear larger makes them even more attractive.

It should be mentioned at this point that Sigmund Freud did come to a remarkably similar conclusion regarding the importance of breasts and mother surrogates in men's sexual life – despite approaching the subject from a very different point of view, including the psycho-analysis of some extraordinary Viennese women.

> No one who has seen a baby sinking back satiated from the breast and falling asleep with flushed cheeks and a blissful smile can escape the reflection that this picture persists as a prototype of the expression of sexual satisfaction in later life ... At a time at which the first beginnings of sexual satisfaction are still linked with the taking of nourishment, the sexual instinct has a sexual object outside the infant's own body in the shape of his mother's breast ... There are thus good reasons why a child sucking at his mother's breast has become the prototype of every relation of love.

A sideways glance at human sexual behaviour illustrates the pivotal function that breasts play in sexual gratification. The classic Kinsey investigation into the sexual behaviour of Americans found that 98 per cent of sex between husbands and wives involved manual stimu-lation of the breasts, while 93 per cent involved mouth-to-breast contact. With the evolution of permanent breasts, sex for men reached new levels of gratification, combining elements of maternal comfort with orgasmic frisson – for the first time 'love', an emotional and physical dependency, became an integral part of our species' sexual behaviour. Men could now simultaneously indulge their infan-tile desires by suckling at the nipple while having an orgasm.

The role of females in promoting the emergence of infantile behav-iour is strengthened by the physiological feedback mechanisms that exist to provide sexual pleasure from breast stimulation during lovemaking. Evolution, of course, does not instil a sense of knowledge into individuals, but rather operates by providing them with incentives. In this situation, it would have favoured females who derived sexual pleasure from having their breasts stimulated, and who would therefore have solicited infantile behaviour from males to provide this. The human ethologist Irenäus Eibl-Ebesfeldt has also noted the dual role of the breast:

> The close relationship between infantile care and sexual behaviour is manifested in the woman in many ways. During sexual arousal, as in nursing, the nipples are erect and can lactate. In some women stimulation of the nipples suffices to elicit an orgasm. Finally, uterine contractions can occur both during nursing and sexual intercourse. This is consistent with the fact that the breast is not only a source of nourishment for the infant but is a sexual releaser for the male.

Infantile sucking has become an integral part of male and female sexual activity. Much as babies suck at virtually everything in the absence of the breast, from clothing to dummies, even elbows pushed up against their mouths, so adults too, in their state of rediscovering and reliving infantile pleasures, are compelled to suck at various parts of their partner's body during lovemaking. Frequently, this involves merely sucking at the skin, a behaviour that frequently causes intense embarrassment for teenagers when it happens on the neck and results in highly visible 'love bites'. However, there is one entire structure that most probably owes its origin entirely to this particular infantile behaviour. The ear lobes. While all our primate relatives have ears, theirs are far firmer structures and curve straight around to the head at their base. Human ears are unique in possessing an extra pendulous structure at their base. Early anatomists dismissed them as function-less, and described them as 'a new feature which apparently serves no useful purpose, unless it is pierced for the carrying of ornaments', or as 'useless, fatty excrescences'. Yet we know that ear lobes are often involved during lovemaking, and that during intense sexual arousal they become swollen and engorged with blood. The Kinsey report even reported that some people can reach an orgasm simply by stim-ulating these tiny structures. They are, without doubt, erotic and sexual in their origin. They are also, minor variations notwithstanding, just about identical in shape to nipples and located in the ideal posi-tion to be sucked during lovemaking.

As for the occurrence of infantile behaviours in both sexes – such as sucking and kissing, in addition to strong feelings of childlike depend-ency on their sexual partner – the reason for this is straightforward. By choosing to mate with males who displayed infantile features, the ancestral females pulled the whole species, themselves included, into a more infantile condition – since daughters as well as sons would have inherited their father's infantile qualities. The situation is no different from that in which a woman who marries a very tall man is as likely to give birth to tall daughters as tall sons. While women have strong maternal instincts – which, for reasons mentioned earlier, can be directed towards their sexual partner – they have also become consid-erably infantised themselves and therefore possess a great degree of infantile dependence too. This 'sexes sharing genes' phenomenon explains why women are so often torn between wanting to mother their partner, and be looked after by them too. It is a dilemma about which psychoanalyst Nancy Chodorow commented:

As a result of being parented by a woman, both sexes are looking for a return to this emotional and physical union. A man
achieves this directly through the heterosexual bond which
replicates for him emotionally the early mother–infant exclusiv-
ity which he seeks to recreate. He is supported in this endeavor
by women, who, through their own development, have remained
open to the relational needs, have retained an ongoing inner
affective life, and have learned to deny the limitations of masculine lovers for both psychological and practical reasons.

The desire of both sexes to find sexual partners who can take over
from their own parents – as well as look after their children – is
supported by some recent work that has looked at the physical qualities of our partners. Next time you have the chance, perhaps over a
meal or even while watching television, take a good long look at
your partner's face. Does it by any chance remind you of anyone?
Psychologists at St Andrews University in Scotland photographed a
number of volunteers' faces, and manipulated the images on a
computer so that they appeared to belong to the opposite sex. They
then showed them back to their original owners for their opinion.
So, what did they think? For a start, none of the volunteers recognised the manipulated faces as being derived from their own. Most
importantly, however, they found these sex-changed images of
themselves extremely attractive.

But were the volunteers attracted to these images simply because
they found themselves attractive, or was it because they reminded them
of their parent of the opposite sex? A strong indication that it was the
latter came from a further test that involved showing the volunteers a
series of photographs of a single face of the opposite sex – one that was

entirely unrelated to their own – that had been computer-manipulated so that it appeared to belong to people of different ages. The volunteers were then asked to score each photograph within the age series according to its attractiveness. It turned out that volunteers whose parents were older than thirty when they were born, consistently found older faces more attractive than volunteers whose parents were younger than thirty. The more the face resembled that of their parents, therefore, the more attractive it became. To further prove the point, researchers are now testing whether we are attracted to people who have the same eye and hair colour as ourselves, or the same as our parent of the opposite sex. The preliminary findings indicate that the latter is true. It therefore seems that, in general, men choose partners who resemble their mother, while women choose partners who resemble their father. Freud would undoubtedly concur.

THE DIVINE MOTHER

Deep in the hills of the Haute-Garonne in the south of France, a human – we do not know if it was a man or a woman – must have spent many hours or even days whittling away at a piece of mammoth ivory to create the most exquisite female statuette. As to why this figurine was carved, nobody can be absolutely sure, but her enormous pendulous breasts that droop so low that they rest gently on her hips, as well as her even more massive buttocks, suggest that she symbolised exaggerated fertility and was one of the early mother goddesses. She is the 'Venus of Lespugue' who, along with a multitude of other female figurines that have been carved over the last 30,000 years, provides further compelling evidence for the power and ancient history of mother-love.

The majority of the figurines that have been discovered date from 25,000 to 12,000 years ago and show women in various states of age and reproductive condition, including a few in the process of giving birth. A somewhat audacious French gynaecologist, Jean-Pierre Duhard, has even gone so far as to suggest that these figurines represent a series of studies of female pregnancy. He furthermore claims to be able to tell not only whether each woman is pregnant, but also whether it is her first pregnancy, and how many children she has already had. Other scientists have been less adventurous and have merely classified the women into various age groups. Out of 132 female figurines, one such study concluded that thirty were too young to have had a child, twenty-three were middle-aged and pregnant, fifty were middle-aged but not pregnant, and twenty-nine were sagging all over and definitely old.

But these details must not be allowed to detract from the single most astonishing fact about these little statuettes. Far from finding miniature depictions of relic Rambos or Stone Age Schwarzeneggers, as one might reasonably expect to find in a society that undoubtedly had to fight for its survival, the vast majority of the human-like carvings that have been found are of women. If these little carvings were simply mementos of everyday life, we would certainly have expected them to represent the whole gamut of life, including men and children. But they were obviously not the prehistoric equivalent of wallet photos, and the vast preponderance of females, and pregnant females at that, clearly shows that women held some very special place in these societies. As for their use, because of their handy size most of these statuettes were probably carried around as amulets to protect their bearers from evil, or as lucky charms to bring good luck. Alternatively, because some were carved so that the woman's legs formed a spike, they may have been put in the ground to form shrines. The message, however, could not be more clear: as soon as art emerged, it was used to celebrate the role of the mother. The evidence from these early carvings, as well as from later paintings, frescos and pottery, points directly towards the Stone Age societies being female-fixated. From around 30,000 to around 500 BC, when the last goddess temple closed, motherhood was divine. As soon as painting and pottery emerged, women dominated the religious imagery throughout Europe and the Middle East. It was a time when Mother Nature was seen as being omnipotent, and where the mother goddess was as much feared for her capricious temper as she was adored for her benevolence. Describing the traditional goddess, the classical historian Donald Mackenzie wrote:

> She was the Fate who measured the lives of men, who sent disasters as well as blessings, and was associated with lions and snakes as well as doves and deer. Withal, she was a voluptuous wanton. Like the Babylonian Ishtar, who was the lover of Gilgamish in one hour and his unrelenting enemy the next, she was fickle and changeable as the wind and the seasons.

Even today, in our male-dominated world, the phrase 'Father Nature' seems incongruous and self-contradictory.

Perhaps the most amazing example of the dominance of the mother goddess in these early civilisations comes from the world's first ever city, called Çatal Hüyük, in central Turkey, which existed from 6500 BC until 5500 BC. Surrounded by peasant villages with their primitive mud-huts, it almost belonged to another epoch from the rest of the world, having sun-dried brick houses that could only be entered through holes in the roof, as well as streets, courtyards and temples that spread over more than thirty-five acres. No doorways have ever been found. But perhaps most surprising of all was not the style of building, but rather the religion that pervaded every aspect of Çatal Hüyük. In one acre alone, as many as forty shrines were excavated – one for every four to five homes – all of which were found to be dedicated to a mother goddess. Within these shrines, there are numerous paintings and reliefs that depict the mother goddess in many forms – young, pregnant and old – as well as extraordinary features such as clay breasts, heavy as if full of milk, that were sculpted into the walls of the shrine and which clearly showed them to be places where a goddess was worshipped. Interestingly, these female deities even occasionally bore a distinct resemblance to the Venuses that had been carved more than 20,000 years earlier. It is, of course, no less than we should expect of a society in which men relied on women – either in the form of their mother or their wife – for security, succour and support.

But Çatal Hüyük provides one further tantalising piece of evidence that points directly towards the fact that as long ago as 6000 BC women might have been fully aware of the child–mother relationship that existed between husband and wife. This historical gem comes from a house known as E VI, 30, and comes in the form of a greenish slate wall carving. It simply shows the mother goddess as two female bodies, back-to-back, one nursing an infant and the other embracing a lover. The position of the woman is identical in each, it is just the object of their embrace that differs. Even at a distance of 8,000 years, this carving evidently illustrates women's role in providing for the identical needs of infants and men. Moreover, this tableau is not alone in attesting that this dual role of women was common belief. Just to the west of Çatal Hüyük lies another archaeological site called Hacilar, a site that was occupied as long ago as 7000 BC. Among various pottery artefacts have been found a small number of exquisite baked clay statuettes, varying in height between two and six inches tall, showing the goddess with a child like lover. With his body entwined around the great goddess, this diminutive male simultaneously appears as her child, with his mouth resting close to her breast, and also as her lover, with his genitals pressing close to hers. Although the image is very different from that of Çatal Hüyük, the symbolism is just the same – of the essentially similar qualities of child and lover.

So what happened to the mother-goddess cult after 500 BC? The answer is that it became absorbed into another highly successful, but this time male-dominated, cult – one that has just reached its two thousandth birthday. As it swept across the world, Christianity voraciously incorporated the icons, ceremonies and deities of previous religions into its own – an ingenious ploy designed to appease those who still hankered after a previous religion. The date of Christmas, for example, was chosen to coincide with the winter solstice celebrations of the rival heathen religion of Mithra. Meanwhile, the mother goddess lives on in the form of the Virgin Mary. Of all the possible names that the Latin translators of the Bible could have chosen for Jesus's mother – her original Hebrew name being *mem-rech-aleph-mem* – the most likely reason why they chose 'Mary' was because of its origin from the Latin word *mare*, meaning sea, and because the sea was seen as the body of the ancient goddess Isis. Indeed, the Virgin Mary is so inextricably linked to the goddess Isis that she has even inherited the role as patroness of mariners, and so similar are images of Isis suckling the infant Horus to those of the Madonna and Child that they are sometimes mistakenly worshipped by Christians.

As for why religion become male-dominated – almost certainly this reflected a shift in society towards increasing male power. During the hundreds of thousands of years when our species obtained most of its food by gathering or hunting, it would have been extremely difficult for males to hold power over females by monopolising any one food-source. However, with the arrival of a pastoral or agricultural way of life, for the first time males could have held the key to survival. From this point onwards, it was just a matter of time before men insisted that the sexual politics of the gods both mirrored and justified the male-dominated situation on earth. According to the Austrian psychoanalyst Melanie Klein, a major feature of the masculine personality is a fear of dependency on a woman, and the acquisition of power by males with the dawn of agriculture would finally have allowed this to be expressed. Klein saw men as being motivated to dominate women in order to cope with their own dependency needs. In some societies, there is a more open expression of female resentment among males. In the Hopi Indians, for example, anthropologist Abie Schlegel found that beneath the superficial image of women being prized as the source of life, runs a deep resentment among the men for the absolute power that they hold early in life. It is interesting to note that many early anthropologists, such as J.J. Bachofen and Sir James Frazer, as well as numerous Marxists, including Friedrich Engels, have suggested that all societies have to pass through a female-dominated phase, in order to subdue men's animal lusts. While truly female-dominated societies – like those of the mythical Amazons in which all males were subjugated to domestic drudgery – are very unlikely to have existed, male social and spiritual dominance is unquestionably a recent arrival in the long history of our species.

Before finally leaving the subject of our forebears' devotion to divine mother figures, there is one last intriguing possibility regarding the origin of the most important Christian symbol of all – the cross. Could the power of the ancient mother goddess be so great that she has managed not only to infiltrate the fabric of the Christian faith, but even lurk clandestinely behind its defining icon? The facts suggest that this might actually be the case, since crosses were popular in ancient Cyprus in the Chalcolithic period between 4000 BC and 2500 BC. But what did the cross represent before Christians started to use it? In a fascinating series of steps, archaeologists have shown how detailed statues of female figures with their arms outstretched and knees bent up, gradually became ever more stylised until they ended up as a simple cross – the vertical line representing the female's body and the horizontal line her outstretched arms. Are Christians, therefore, bearing ancient depictions of a goddess, knees up and straining as she gives birth, around their necks? Is there no end to the mother goddess's grip on our species?

THE BABY-FACE

So much for signs that women's bodies and behaviour have been moulded to satisfy men's craving for both a mother figure and sex object, but are there any tell-tale aspects of men's bodies that reveal a preference for infantile males by many generations of women? Clearly, there is a vast amount of variation in terms of what women find attractive in men – if this was not the case the vast majority of human males would find themselves in love's reject bin. But there is also considerable agreement – both within and between cultures – on the type of man that women find attractive, especially in terms of his physical appearance.

Psychologist Michael Cunningham and his colleagues at the University of Louisville set out to investigate an apparent dilemma regarding what women look for in their partner:

On the one hand, a woman may desire a man who appears dominant and mature, so that he might successfully compete against other men and provide ecological and genetic resources to help ensure her survival and that of her children. On the other hand, a woman might not only desire her mate to protect her but she may also wish to cooperate with and even protect and nurture her mate ... women report that they desire men who are tender, gentle, sensitive, kind, and understanding. A man who looks too mature and too powerful, then, may not arouse the woman's warm, caregiving feelings and may not elicit as much attraction as a man who can stimulate nurturant responses.

As Cunningham discovered, the most attractive human male face turns out to be an intriguing mixture of mature and what he calls 'babyish' features. While a certain degree of maturity is considered attractive in men, features that are overly masculine – such as very thick eyebrows, small eyes, thin lips and square jaws – are not considered particularly handsome. In contrast, infantile features including big eyes, higher set eyebrows and narrower chins have been linked to greater attractiveness. The ideal male face, according to this research, is one that provides the right balance of mature features, to demonstrate a male's ability to provide protection, and childlike features that elicit nurturing behaviour on the part of the female.

In another study, David Perrett, a psychologist from the University of St Andrews, photographed a number of men and women, and then computer-manipulated each face to create a series of further photographs in which the subject was made to look more or less feminine in the case of women, or more or less masculine in the case of men. Perrett and his team carried out the same procedure in both Scotland and Japan. When men were shown the series of women's photographs, they found, as predicted, the 'super-feminine' versions more attractive than the actual and 'reduced-feminine' versions. However, when women were shown the series of photographs of men, he discovered that they found the 'reduced-masculine' versions more attractive than both the actual and 'super-masculine' versions. Perrett noted, 'Enhancing masculine facial characteristics increased both perceived dominance and negative attributions (for example, coldness or dishonesty) relevant to relationships and paternal invest- ment.' He concluded that sexual selection by females therefore reduces the development of highly masculine features in males, and promotes the infantising of our species.

Moreover, the findings of sociologists Alan Booth and James Dabbs suggest that there is a highly tangible benefit to choosing immature partners. More precisely, according to them, women today have as just as much reason to favour men who retain immature levels of sex hormones as they ever did – especially if they want to avoid an ill-tempered and short-lived marriage. In a study that involved meas- uring the testosterone levels of 4,462 men, and comparing these levels with the quality of the men's marriages, Booth and Dabbs came to the following conclusion:

Though aggression and dominance behavior are well suited to gathering and amassing resources, and achieving and maintaining

status, when unchecked they are not conducive to the cooperation and mutual support essential to intimate heterosexual relations, especially those of an egalitarian nature … Our analysis
shows that men with higher testosterone production report
being less likely to marry and more likely to divorce. Once
married this type of individual is more likely to leave home
because he is having trouble getting along with his wife, more
likely to have extramarital sex, more likely to report hitting or
throwing things at his wife, and more likely to experience a lower
quality of spousal interaction.

Recent research by Peter Gray and a team from Harvard University
has also found that married men who spend time with their wives and
children have lower levels of testosterone than bachelors. The
researchers suggest that lower levels of testosterone make fathers less
likely to stray and therefore devote more of their energy to the family
rather than looking for another partner. In other words, it seems that
women face a stark choice. Pair up with a man who is oozing with
testosterone – and has all the mature physical characteristics to prove
it – and risk a higher chance of physical abuse, infidelity and a short,
unhappy marriage. Alternatively, choose a more immature-looking
partner and opt for an increased chance of a longer, more caring
marriage. So why do some women still favour macho- and testos-
terone-laden men as their sexual partners? The answer is because
these men can have something to offer in terms of benefits relating
to their high status and their potentially greater ability to successfully
compete for essential resources. However, while some women are
clearly willing to risk a tempestuous relationship for high short-term
gains, this is clearly an extremely risky strategy that can easily lead to

desertion, hence the more widespread attraction of women to faithful, caring and immature partners.

Men, quite naturally, have hardly remained passive in the face of women's strong incentive to choose immature-looking and behaving partners. In terms of their behaviour, young men are far more likely to engage in boisterous games and joking sessions with other men when displaying to members of the opposite sex, than embarking on aggressive displays of dominance, as would chimpanzee males. Whereas chimpanzee males attack, bite and beat each other up, human males play, giggle and smile. But there is another quite extraordinary behaviour that human males have recently adopted in order to exaggerate their immature appearance and thus make themselves super-attractive to females. This is a behaviour that is objectively so bizarre that it brings to mind Bob Newhart's imaginary account of introducing tobacco to civilisation for its almost unimaginable absurdity. In this famous sketch, Newhart recounts a conversation is between Sir Walter Raleigh (aka 'Nutty Walt') and the head of the West Indies Company in England.

Picture the scene. The head of the West Indies company reclines in his plush leather chair as he presses the anachronistic telephone to his ear. At the other end of the telephone line is Sir Walter Raleigh – Nutty Walt – the company's eccentric and incorrigible globetrotter, who is calling from America to extol the virtue of his latest discovery. Having just received a boatload of wild turkeys – that are, incidentally, still running uncontrollably around London – the company boss eagerly waits to hear what Nutty Walt intends to unleash on London's unsuspecting population this time. He is told that it is a plant called 'To-ba-cco', and that Nutty Walt's next shipment is to be packed to the gunnels with dried tobacco leaves. Despite agreeing that England is already up to its ears with dry leaves, Nutty Walt insists that these particular dried leaves are something really special.

What follows, however, is so improbable – even by Nutty Walt's standards – that the company boss breaks down into raptures of laughter. Firstly he hears that, when ground down, dried tobacco leaves can be made into a substance called 'snuff' – a powder that, when shoved up one's nose, has the effect of causing uncontrollable sneezing. But then for the pièce de resistance. Nutty Walt goes on to describe the most exciting, as well as assuredly popular, use of this wonder-leaf. With great enthusiasm he describes how it is possible to shred the dried tobacco leaves, roll them in paper, place them between one's lips, and set fire to them! Desperately trying to control his mirth, the boss respectfully suggests that standing in front of the fireplace might provide much the same experience, and that it could be a little difficult selling the public the idea of sticking burning leaves into their mouths. In fact, ever since Nutty Walt put his cape down over the mud, they have been a little worried about him…

Now imagine the same conversation, with humble deference to Bob Newhart's original sketch, except that Sir Walter Raleigh is trying to describe the latest fashion craze to hit previously bearded America.

'What is it this time Walt? You got another winner for us this time do you? Sha-ving – what's shaving, Walt? It's the latest fashion, is it? And you think it looks really great.

'It involves a very sharp knife … and lots of foam? First of all you spread the foam all over your face with a brush. What do you do then, Walt? You make the knife so sharp that it can cut off your nose with a single swipe … you then bring the knife towards your face and you scrape it all over so that all the hair falls off? You know, Walt, we've got skin diseases that do that pretty well already.. ...

'Then what do you do, Walt? You wash your face, and then sprinkle alcohol all over it so that it stings, but that stops the blood flowing from the cuts you've just made ... and they you put sticking plasters over any cuts that just won't stop bleeding? You know what, Walt, it seems you can put your head into a lawnmower and have the same thing going for you, you know. You see, Walt, we've been a little worried about you, you know ... ever since you put your cape down over that mud. You see, Walt, I think you're going to have kind of a tough time selling people on cutting their faces with knives ...'

But still men shave to please the baby-seekers, since the beard is perhaps the human male's most conspicuous indicator of maturity. By the time they are sixty, men will have spent around 2,555 hours performing this strange ritual, and cut off about twenty feet of beard hair. As to why human males have facial hair in the first place, this is something of a mystery, although one reason is that it initially served to exaggerate the already heavier jaw of the human male, a feature that gives him a more masculine appearance. Alternitively it could simply be an unavoidable consequence of our infantile origins. As mentioned earlier, even unborn chimpanzees that have yet to develop hair over most of thier body possess facial hair around their mouth. To accentuate the juvenile effect of shaving, most Western men also cut their hair short which, when combined with a shaved face, regresses their facial hair back to a state that would naturally occur only in a baby. What this incidentally creates is a face that looks decidedly androgynous – something that is even more pronounced in males with small infantile noses and large eyes – because of the almost total removal of male sexual features, other than perhaps a slightly heavier jaw. Some boy bands, the heart-throbs of teenagers in the late nineties, took this effect to its furthest extreme by shaving every conspicuous hair on their bodies. Contrary to the saying that girls 'like boys who look like girls', a more accurate description might be that girls 'like boys who look like babies'.

THE LOVE REVOLUTION

What better bond than a child's dependency on its mother for females to exploit at a time when they needed to coerce males into helping to raise their offspring. Two million years ago, when Africa was experiencing massive droughts, female sexual selection needed to kick into action like never before to transform our promiscuous ancestral males into diligent attentive partners. As far as females were concerned, any strategy that managed to get males to provide them and their offspring with food and protection would have been favoured, and of all the measures to achieve this – including synchro- nised menstrual cycles, concealed ovulation and permanent sexual receptivity – that of selecting infantile males, who retained a strong dependence on a mother figure, was to prove by far the most success- ful. Indeed, the males of two million years ago were the ideal raw material for this process as a result of the significant infantising that had occurred to our ancestors by this time. Because of the advantages of infantile group behaviour, these males would already have formed far stronger relationships with their mother than ever before – just like bonobos today. Without this first phase of infantising – which dated back to the moment that our earliest ancestors left the trees – there would not have been the raw material from which females could select the most extreme infantile males of all. Essentially, it primed our species for the next phase of infantising which led to the evolution of infantile pair-bonding. The whole situation played right into the hands of the females. With the emergence of the male who fled directly from the protective bosom of their mother to that of their wife, sexual love had finally evolved. While the mother–child relationship fundamentally underpins the human pair-bond, count-less generations of females selecting infantile males has caused the whole species – females included – to become infantised, since daughters as well as sons have inherited their father's characteristics. The situation that our species finds itself in today, therefore, is that of both sexes having to act as a pseudo-parent for their partner, while simultaneously wanting to be parented themselves.

A minor step, one might imagine, in the evolution of a species – merely a slight shift in the relationship between the two sexes. Such a conclusion could not be more wrong. It was like opening the flood-gates for our species. Indeed, the metaphorical gates at the boundary of Africa were literally opened at this time, as our new ancestors flooded into the rest of the world. And why? The reason is that this simple change in the relationship between males and females totally revolutionised the way that our ancestors interacted with everything around them. The emergence of increasingly infantile and athletic bodies, combined with a division of labour between the sexes – all consequences of the infantising process – meant that they were superbly adapted to exploiting the world as never before. At last, the critial threshold had been crossed – from ape-like to unquestionably human-like. Love had revolutionised the world.

What is arguably most impressive about the infantising process is that females nearly pulled off this magnificent trick. They nearly subjugated men to a lifelong servile position that would have suited them so perfectly. Human males do, to a large extent, look for a seamless transition from mother to wife. As psychologists Evelyn Pitcher and Lynn Schultz describe the present situation:

> The male, vulnerable to erotic seduction, can join in the relationship the female craves, and in doing so can satisfy his own basic need to return to the female-mother. Often he is held in this relationship by well-developed techniques of nurturing and appeals to appropriate rules of conduct. Nurturance and reprimand will later be summoned by the female in her control of her children and passed on as fundamental modes of behavior to her daughters.

There was, however, one tiny fly in the female's male-taming ointment. Men. Men, like women, also have agendas – and these are not the same as those of females. Given half a chance, there is very little doubt that human males and females would choose extremely different sexual arrangements. For a woman, the ideal situation would naturally involve having a faithful husband who devoted his entire life to providing for her offspring – regardless of whether her children were actually fathered by him or not. For a man, on the other hand, who can father literally hundreds of children in a single year, the ideal situation is very different. For him, the perfect scenario is more likely to involve the mother doing all the child-rearing, while he puts maximum effort into inseminating as many women as possible. Yet despite this male ideal, females have been remarkably successful in using their selective mating powers to create the ideal father for their children. The vast majority of men do fall in love, and form strong emotional bonds with a single woman at a time. And it is all down to women – thousands of generations of women – who have chosen infantile, loving, dependent and attentive men to be at the receiving end of their maternal as well as sexual favours. Well, most of their sexual favours anyway.

So, to return to the original point posed at the beginning of this chapter. What is the natural mating system of the human species? The answer is that we are essentially a promiscuous species – or at least we would be, if only we could develop beyond the immature stage of needing the stability of a mother or father figure throughout our lives. It is this partner dependence that gives our species the superficial appearance of being monogamous. Sometimes, with or without society's blessing, we indulge in promiscuity. We dabble in being maturely promiscuous. However, like a child taking its first tentative steps away from its parent, we simultaneously crave the emotional support of a long-term partner – our parent substitute.

chapter six

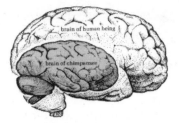

brain of human being

brain of chimpanzee

the
accidental
brain

The following was contained in a letter that was sent to the child psychologist David Henry Feldman:

> [My child] conceived a passion for languages which lasted from about 22 months to age four [studying Hebrew, classical Greek, French, Italian, Russian, Spanish, Yiddish, German, Egyptian hieroglyphics and Sanskrit] … Shortly after he turned four, I noted that he had shown little interest in the material which his tutors selected for him. Instead, he was interested in pursuing the [personal] relationship which he had with each of them. I told him that if he was no longer interested in the language that was all right, but that meant we would stop the lessons and retain the personal relationships with the tutors as friends. He said that he

was not really interested any more because he had figured out the answer to the question. Ever the straight man, I asked him, 'What question?' 'Oh,' he said, looking at me with helpful attention. 'I've figured out that there was a parent language for these languages (he listed the Indo-European ones) but not for these. It had 11 cases.' I told him that scholars called the parent 'Indo-European' and that I had been taught in college that it had ten cases. He said, 'Scholars can be wrong. It had 11 cases!'

From an incredibly early age, the power of the human mind is awe-inspiring – and never more so than in the case of child prodigies. Without the benefit of vast amounts of experience, such inchoate minds are capable of the most extraordinary feats – from grasping complex mathematical concepts to composing intricate symphonies. Of course, such minds are merely ahead of their time, and human minds of any age are truly astonishing, especially since they seem to spring out of biological nowhere. We stand so far apart from all other animals in our mental abilities that comparisons seem banal. Chimpanzees, for example, take years to master the simple task of smashing a nut between two rocks. So how did our species acquire such a powerful super-computer? How did this 'meat machine' – a phrase coined by Marvin Minsky of the Massachusetts Institute of Technology – become so incredibly magnificent in both its ability and size?

Going back in time, the facts regarding the explosion of the human brain speak for themselves. Before 2.5 million years ago, our ancestors' brains were little different in size from those of chimpanzees. By 1.8 million years they were about twice the size of a chimpanzee's – that is, about half the size of modern humans'. By one million years ago, they were about two-thirds the size of ours today, and by 0.4 million years they had reached their present-day size of about 1.4kg. Ever since just over two million years ago, the human brain has been getting larger and larger. Why?

The first clue comes from our earliest days, and the fact that the human animal is incredibly vulnerable and helpless when it is first born.

THE MAGGOT BABY

The human brain has received a considerable amount of scurrilous press in scientific and popular literature. According to common belief, our species' enormous brain is unreservedly guilty of putting our very survival at risk during our first few months of life, by growing so rapidly in the womb that it forces us to be ejected long before our defenceless little body is ready to face the world. It is accused of giving our mother no alternative but to expel her beloved baby from her womb prematurely, before it makes the baby's head too big to escape through its mother's birth canal – a situation that would condemn both mother and child to a horrible death. But is the brain really guilty of this heinous crime? Is the brain's excessive rate of growth genuinely responsible for our pathetically immature state when we are born?

To start with, difficult births are not unique to humans but rather typical of primates in general. All primates, from tiny monkeys to chimpanzees and humans, have relatively large heads for the size of their bodies. Human babies are actually born with slightly smaller heads, compared with the size of the rest of the body, than some other primates, including orang-utans and gibbons. In certain smaller primates, the problem of giving birth to a baby with a large head can be particularly acute. In squirrel monkeys, for example, giving birth is often so difficult that the babies themselves have to help with their own birth. While they are being born, baby squirrel monkeys frequently resort to grabbing the fur of their mother with their hands in a desperate attempt to haul themselves out of her body. Sadly, despite these measures, about half of all squirrel monkey births in zoos are unsuccessful and end up with the baby dying. In comparison, we humans seem to have little to complain about.

Returning to the woefully undeveloped state of the human baby when it is born, it is simply not true that the brain grows so fast that it 'forces' the baby to be born before it is ready. The human brain grows at exactly the same rate as the brains of all other primates. The vast majority of the size difference between human brains and those of other primates results from a far longer period of growth after the baby has been born, not before. Whereas chimpanzee brains roughly double in size between birth and adulthood, and have largely finished growing by a year after birth, human brains nearly quadruple in size and only finally stop growing when we reach twenty years old. Deborah Yurgelun-Todd and others at the Harvard Medical School, found that young teenagers consistently fail to correctly identify certain facial expressions – mistaking, for example, a fear expression for sadness or anger – and also use their brains' frontal lobes far less than adults when looking at different faces. On this basis, Yurgelun-Todd suggests that the human brain takes until the late teens to become fully developed in our species. As regards the 'premature ejection' belief, the length of the human pregnancy is actually four weeks longer than that of chimpanzees and gorillas, so if anything human babies' bodies should be more developed.

If it is not a fast-growing brain, or an abbreviated gestation, then what does explain why we are born so undeveloped? The answer is that the human body develops at an incredibly slow rate – it takes far longer to progress from one developmental stage to the next than in any other primate. Despite the fact that we put on weight as fast as other primates in the womb, we nevertheless fail to mature at anything like the same rate. In fact, if human babies were to stay in their mother's womb until they had reached the same stage of development that chimpanzee babies are in when they are born, then they would have to stay inside their mothers for nearly two years. By this time, the baby would weigh 10kg, and the mother would need hips the dimensions of a toilet seat to give birth to such a monster. As Wilton M. Krogman, anthropologist and expert on child growth and development, noted: 'Man has absolutely the most protracted period of infancy, childhood and juvenility of all forms of life ... Nearly thirty percent of his entire life-span is devoted to growing.'

Everything about the human body is staggeringly underdeveloped when it is born. The ends of the long bones that form our arms and legs, as well as the bones of our fingers and toes, are still soft and cartilaginous, rather than being strengthened with bone as in other primates. As a result, the human baby has to be well over a year old before it can even begin to withstand the boisterous rough and tumble of the real world. In any other species, this situation would be lethal – the slightest drop or fall would result in a massive amount of damage to an animal such as a chimpanzee. Our skulls, too, are fragile almost beyond belief at birth. If other newborn primates have heads as tough as footballs, we have heads as delicate as watermelons. Whereas the bones that make up chimpanzee skulls have fused by the time they reach three years old, in humans these bones usually take twenty-five to thirty years to fuse completely. Newborn babies even have massive holes between their flexible skull bones, allowing their brains to be prodded without any protection whatsoever. It is an astonishingly dangerous state to be born in. As all mothers know, the largest of these holes is the anterior fontanelle, or 'soft spot', which takes about two and a half years to close completely. It seems that, as great as the benefits of our hugely extended period of development may be, it has undoubtedly made us incredibly fragile, delicate and vulnerable during our early years.

Because of their slow rate of development, humans have no choice but to devote more care and attention to their newborn young than any other animal on this planet. Whereas other newborn primate babies can at least hold on to their mother, human babies can merely wriggle around like helpless maggots. Similarly, whereas two-week-old chimpanzee infants can raise their heads and chests by extending their arms when lying face down, human babies do not reach this level of development until they are between three and five months old. Chimpanzee infants can also hold their heads steady when sitting upright from virtually the day that they are born, while it takes roughly four months for human babies to be able to achieve this. As Austin Riesen and Elaine Kinder, working at the Yerkes Laboratory of Primate Biology, describe:

> Within a few hours of birth, when the [chimpanzee] infant is in vertical or prone [face-down] positions, head movements associated with mouthing normally become quite vigorous ... This is in striking contrast to the human infant which when in the prone position has so limited range of movement of the head that it can only momentarily raise and turn it sufficiently to free the nose.

In short, human mothers have had to cope with giving birth to babies that are so underdeveloped that they should really still be in the womb – and this invariably means providing them with special 'external wombs' in which to keep them until they are sufficiently developed for their true birth into the real world. As mothers bind their babies close to their body with swathes of cloth, they provide them with cosy, warm and dark imitations of the internal womb, so that they can continue their development safely and unabated. And when babies are not being carried in external wombs attached to their mother's body, they are placed in static womb substitutes, such as cots and cradles, or even being tightly swaddled to mimic the pressurised environment of the fluid-filled womb. The most accurate way to look at a Moses basket, cot, or even car seat, is not simply as a comfortable place to sleep, but rather as a vital womb replacement in which babies are allowed to make up for lost time in the mother. We often hear of the need for babies to be close to their mother, while in fact what they could really do with is to be back inside her.

Even after we have left the security of our cot or car seat, our development still progresses at a snail's pace. Our teeth take twice as long to come through – chimpanzee milk teeth erupt at three to four months and their permanent teeth at three years, whereas our milk teeth erupt at about eight months and our permanent teeth at about six years. We keep on growing for longer than chimpanzees – they are fully grown by age eleven, while we take twenty years to become fully grown – and human females become sexually mature far later than chimpanzee females. Even dying takes longer to happen – if we were a chimpanzee, most of us would be dead by the time we were thirty-five years old. All in all, in the words of Stephen Jay Gould, 'Without belaboring the point any further, I think it fair to state that general retardation in human development (relative to other primates) is a fact.'

In conclusion, despite being born when they are the 'right' size – and this goes for the size of the head as much as the rest of the body – humans are staggeringly retarded in their development, and this alone explains why we are so helpless early in life. Considering the size of their bodies, human babies should be much further developed than they are. It seems that the infantising of our species has resulted in each developmental stage lasting for so long that, by the time that it is born, the human body is already far behind in its maturity. Humans, in other words, spend a lot longer in the early phases of growth than any of our close relatives.

How, though, does this fit in with the overall picture of our species becoming infantised to make us more sociable, cooperative and, most recently – under the influence of females – more loving husbands? How does our brain fit into this scheme of things?

IS SHE IMPORTANT?

The entire package does indeed seem to make an enormous amount of sense. So the story goes, on an evolutionary level, the human brain is big because, over successive generations, people with big brains have been 'brainier' and more proficient at solving problems – be they social or technical – hence, the average brain size of the human lineage has steadily increased. On a structural level, the bigger the brain gets, then the more neurons it contains, the more complex it can become, and the better it is able to 'work things out'. Finally, it is argued that on an energetic level, the human brain consumes around 16 per cent of the total amount of energy that our body needs to stay alive – at least when we are lying down – and there is no way that such an expensive organ would have evolved if it had not provided members of our species with highly useful 'brain power'. In other words, ever since just over two million years ago, when our ancestors' brains started to get bigger, it has been getting better and better for our species' intellect. End of story. Or is it? What evidence is there that any of the above, barring the figures for energy-consumption, is actually true?

In the late 1800s and early 1900s it became something of a cottage industry for anatomists to obtain the bodies of recently-deceased eminent men, solely to measure the size of their brains. It even became a matter of competition between scientists as to who would ultimately be found to have the biggest brain, although the winner was clearly presented with something of a challenge if he or she intended to celebrate their victory. Against a European average of 1,300 to 1,400g, the French anatomist and palaeontologist Georges Cuvier held the record for most of the nineteenth century with a brain of 1,830g – that was, until the Russian author Ivan Turgenev raced past with a massive brain of over 2,000g. There are even reports of Oliver Cromwell having a brain weight of 2,231g, and Lord Byron's weighing in at a staggering 2,380g. But the picture was far from clear. The American poet Walt Whitman managed to become highly acclaimed with a meagre 1,282g, while the scientist Franz Josef Gall who ironically worked in the field of brain research, came in at an embarrassing 1,198g. To cap it all, or rather put the final nail in the coffin for those who were looking for proof that a large brain is essential for high intelligence, Anatole France, the winner of the 1921 Nobel Prize for Literature, excelled in the literary world with a paltry 1,017g – roughly half that of his fellow author Turgenev. Since then, the smallest fully-functioning human brain that has so far been found is as small as 762g – that is, a mere 5 per cent heavier than the 725g that has been recorded for a gorilla's brain.

The conclusion is beyond doubt. It is that the case for a rigid link between cleverness and brain size is firmly defeated.

Evidence of high levels of redundancy in the human brain comes from a thankfully small number of tragic, yet quite extraordinary, medical cases in which people have lost large parts of their brain, yet retain all the faculties, including language and manual dexterity, that set our species so far from all others. And above all other examples, there is none more celebrated than that involving a railroad employee working in Vermont in the middle of the nineteenth century.

On Wednesday 13 September 1848, just outside the town of Cavendish in Vermont, a small crew of construction workers were preparing a new rail-bed for the Rutland and Burlington Railroad under the supervision of the company's most capable and experienced foreman, Phineas P. Gage. As he had done countless times before, Gage was preparing to remove a lump of rock from the path of the railway using the blunt end of a 'tamping iron', a 3.3m long and 4cm wide iron bar, to ram explosive down a bore hole. But no amount of experience could have prepared Gage or anyone else on his team for what was about to happen. As the tamping iron hurtled down towards the explosive, and with Gage's body following close behind, it must have hit a small piece of rock, from which a tiny spark flew. Tiny or not, it was enough to ignite the explosive that by now had the full weight of the iron bar bearing down on it. Sharp-end up, the tamping iron was propelled like a rocket upwards towards the sky. It was, alas, a spectacle that Phineas Gage was destined not to enjoy since his head was immediately above the iron bar when the explosion happened. As it skyrocketed up, the sharp point of the bar entered the left side of his face just below his cheekbone. It then travelled behind the socket of Gage's left eye, through the left prefrontal lobe, before exiting through the top of his skull and leaving behind a hole 9cm long and 5cm wide. While Gage reeled backwards, the iron bar continued its journey, now smothered with traces of his brain, until it finally landed twenty-five metres away. According to some accounts, Gage never lost consciousness. With half his left frontal lobe, and possibly some of his right frontal lobe, scattered around him, the rest of his brain was bleeding profusely. The other railway workers loaded Gage into an oxcart and rushed him to a nearby hotel where two local doctors patched up his head as best they could. So what became of Phineas P. Gage from then on? For a start, amazingly, he survived. In fact, he recovered so well that within seven months he was back working again. Despite the removal of massive chunks of his brain, Phineas Gage could still talk normally, control every part of his body, reason, argue, plan and to all appearances act as though nothing had ever happened.

But what is to say that Gage was not simply lucky in losing a part of the brain that happened to be of very little use, and that it remains the case that the vast bulk of our 1,400g is essential for us to achieve our unique mental prowess? To investigate this possibility, we need look no further than people who, for a variety of reasons, have had half of their brain removed. In an operation called 'hemispherec-tomy', half of the cerebrum – that is, the part of our brain that is responsible for understanding, speaking, reading, and voluntary movements of the body – is totally removed. The result, one would imagine given the immense importance usually given to brain growth and brain size for our species, should be to render the patient severely mentally disabled. With only half a brain, they should be dispossessed of most, if not all, the major human faculties, such as conscious thought and language. With this thought in mind, consider the following account, written by psychologist Stan Gooch, of two people who had undergone hemispherectomy:

[The patient, a 47-year-old man referred to as E.C.] ... began suffering from attacks of speechlessness, coupled with seizures of the right arm and right side of the face ... On admission to hospital an extensive soft tumour in the left side of the brain was diagnosed. Full left hemispherectomy was performed. Immediately following the operation the patient showed, as would be expected, paralysis of the body, and severe distortion of his ability both to comprehend the speech of others and to express himself verbally. But 10 weeks after the operation, he could repeat single words on command and occasionally communicate in question and answer. By the sixth post-operative month, when asked, 'Is it snowing outside?', E.C. could smile and reply: 'What do you think I am? A mind reader?' More strikingly still, in the fifth post-operative month E.C. displayed sudden recall of entire old familiar songs such as, 'My Country 'Tis of Thee', 'Home on the Range' and church hymns'.

Miss A had been born prematurely by Caesarian section. From birth onward she suffered from persistent weakness and paralysis of the right side of the body. At the age of nine she began to have epileptic fits every five to seven days. These were accompanied by muscular spasms and spastic seizures on the right side ... Interviewed before the operation, then aged 21, she had a mental age of 10 years and 4 months, and an IQ of 70. This girl was unhappy, depressed and acutely aware of her disabilities. She had no amusements, did not read, and was afraid to mix with other people ... [Then, a] Full hemispherectomy of the left dominant hemisphere was performed ... On the day following the operation Miss A was already speaking normally ... Five months after the operation Miss A had regained 'much use' of her previously incapacitated right limbs ... she now had an active social life, was reading newspapers and books for pleasure, and had developed a lively interest in the cinema, in clothes and her appearance. One year after the operation she was working normally in an office.

It seems, therefore, that even with a brain that is half the normal size, humans can still possess the full complement of the mental skills that so impressively distinguish us from all other animals on this planet. It also suggests that it can be far better for a particular part of the brain to be missing altogether rather than remaining in a damaged or diseased state – thereby allowing another part of the brain, in time, to take over its function.

The most astonishing proof that the human brain is far bigger than it needs to be has, however, been left until last. Professor John Lorber, of the University of Sheffield, has scanned the brains of more than six hundred patients with hydrocephalus – that is, people who have parts of their brain cavity filled with fluid in place of brain tissue. Lorber divided his patients into three categories: those with minimally enlarged ventricles, those with fluid-filled ventricles that filled 50 to 70 per cent of the brain cavity, and the most severe group, in which the ventricles filled 95 per cent of the brain cavity. He then measured the IQs of patients in each group. His results were breathtaking.

There's a young student at this university who has an IQ of 126 who has gained a first-class honours degree in mathematics, and is socially completely normal. And yet the boy has virtually no brain. When we did a brain scan on him we saw that instead of the normal 4.5 centimetre thickness of brain tissue between the ventricles [pockets of fluid within the brain] and the cortical surface, there was just a thin layer of mantle measuring a millimetre or so. His cranium is filled mainly with cerebrospinal fluid.

Among the other people in Lorber's study group who had brain cavities that were 95 per cent filled with fluid – in other words, whose brains were a mere 5 per cent of the normal size – over half these people had IQs greater than 100. People whose brains weighed about 100g rather than the normal 1,500g, actually had higher IQs than the average for people with normal-sized brains.

We are presented with the most extraordinary puzzle. Some normal humans have brains that are three times smaller than those of other normal humans – with no discernable reduction of their mental faculties. Moreover, we now learn that people with brains that occupy a tiny fraction of their skull because of a medical condition can have above-average levels of intelligence. So much for human brain size relentlessly increasing because of its huge benefits for human intelligence. In fact, it seems that most of us have brains that are twice, if not three times, as large as they need to be.

SIMPLE MINDS

If present-day humans benefit so little from having large brains, what difference did the ever larger brains of our ancestors have on their technological skills? To start with, as anthropologist Colin Groves has pointed out, the earliest known tools actually precede brain enlargement. The oldest tools that have so far been found – in places such as the Oldupai gorge in Tanzania, around Lake Turkana in Kenya, and in southern Ethiopia – date as far back as 2.5 million years, and are associated with human ancestors who had brains that were very little, if any, larger than those of chimpanzees. By this time, our ancestors had already invented the 'percussion' method of stone tool manufacture, which involved smashing off flakes from around the edge of a large pebble or stone to make an all-purpose cutting and chopping tool. But it is the subsequent evolution of these tools that is most telling.

If you and I were sitting around a table on which were piled ten stone tools that differed in age by over a million years, I would happily bet that you could not even begin to arrange them in order of age. Study them as hard as you like, the chances of finding the slightest increase in sophistication is virtually zero. The reason is that from 1.5 million years to a mere 150,000 years ago the stone tools that our ancestors made hardly changed in any way. It took nearly 1.5 million years for them to progress from hitting flakes off a large stone to give it a cutting edge, to an improved method that involved preparing a stone so that the last hit created a flake of just the right shape and with an incredibly sharp edge. Likewise, it took over two million years, and a doubling of brain size, for our ancestors to manage to attach a sharpened piece of rock to a stick to make a stone-tipped spear or arrow. Tipped spears were most probably invented as recently as 40,000 years ago, while the first certain arrow tips that have been discovered come from northern Germany and date from between 13,000 and 10,000 years ago. And yet, despite these incredibly long periods of negligible innovation, our ancestors' brains relentlessly kept on growing and growing and growing.

Far from simply revealing our ancestors' technological ineptitude, these immensely long periods of creative stagnation may well reveal a great deal about the mental abilities of these people. The psychologist Geoffrey Miller has recently published a long and erudite eulogy to the human brain in which he describes it as a highly sophisticated entertainment centre – one that evolved to woo, amuse, entertain and impress the opposite sex. He describes how our brain growth is the result of both men and women selecting the wittiest and most articulate mates for hundreds of thousands of years, and calls this the Scheherazade effect after an Arabian storyteller who prevented the Sultan from murdering her by keeping him amused with a thousand and one tales. Other scientists, meanwhile, have explained our massive brains along more prosaic and unromantic lines, emphasising the importance of gossip in maintaining social contact in an ever more complex society.

There is, however, a major problem with theories that use language to explain our inordinately large brain, which is that our ancestors' brain expansion predates the emergence of language by between 1.5 and 2 million years. Whereas brain expansion began over 2 million years ago, many anatomical structures that are essential for speech were not in place until 250,000–500,000 years ago, and according to archaeologists such as Richard Klein of Stanford University, actual language – involving syntax and grammar – is likely to be amere 50,000 years old. One reason for thinking that language emerged this late in our evolution concerns a link between the brain's control of the hands and the mouth, and the incredibly recent arrival of complex tools. An association between the hands and the mouth is something that nearly all of us will have experienced at some time or another. I myself, for example, have an irrepressible tendency to move my tongue when I am undertaking some delicate task with my hands, while my daughter performs the most amazing contortions of her lips while playing the violin. As long ago as 1872, in *The Expression of the Emotions in Man and Animals*, Charles Darwin had noted the exact same phenomenon:

> [Persons] cutting anything with a pair of scissors may be seen to move their jaws simultaneously with the blades of the scissors. Children learning to write often twist about their tongues as their fingers move, in a ridiculous fashion.

Applied to the language skills of our ancestors, the logic is that if the brain cannot coordinate the making of complex tools by the hands – such as a spear consisting of two different materials – then it cannot control the making of complex sentences by the mouth. In terms of creating novelty, planning a new sequence of tool-manufacture is essentially no different from planning a new sequence of words to form a sentence. Not require that William Calvin from the University of Washington calls 'spinning a scenario connecting the past with the future'. The late emergence of even the simplest composite tools therefore strongly indicates a similarly late emergence of language. But perhaps most important, is the fact that language requires symbolism and there is no evidence for this until 40,000– 50,000 years ago when ritual burials and art first appear in the archaeological record. Before 50,000 years ago the world was an artistic desert – afterwards symbolism appeared in a fabulous variety of different forms. Until this revolution our ancestors may have produced a wide range of sophisticated grunts, but they were very different from the embedded sentences that we produce today. Unfortunately for Miller's argument, the unavoidable conclusion is that the brain had already reached its full size well before language first appeared in our species. For the vast majority of their evolution, despite having big brains, our ancestors had remarkably simple minds.

THE MICKEY MOUSE SPECIES

If it has not been an incessant demand for better creative language or social skills, what could possibly have caused the staggering enlargement of our brains over the past two million years? To find the answer, all we have to do is take another look at our own bodies. What we see in the mirror every day is a huge head that is perched on the top of a relatively small body. It is not the overall shape of an adult ape, but that of a baby ape. Along with the rest of the infantile package, including a flat face, upright stance, long legs and hairless body, has come a bulbous baby-like head. Far from requiring a separate explanation for its existence, our large brain can be accounted for simply as one more incidental by-product of our species having been infantised. As a result of our hugely extended early developmental stages – as evidenced by our extremely underdeveloped condition when we are born – our brain carries on growing for that much longer than in all other apes. Moreover, this explanation is entirely in keeping both with the lack of evidence that an enlarging brain alone provided any significant benefit to our species for much of its evolution, as well as the finding that our present-day mental faculties are largely independent of brain size.

For over fifty years scientists have noticed that the infantising of a species can lead to increased brain size. It is a phenomenon that has been found in a variety of animals ranging from Central and South American margay cats to the shrew-like African tenrecs. Among the thirty-four species of tenrec, for example, John Eisenberg from the Smithsonian Institute discovered that Talazac's tenrec not only had the most extended development, but also had the greatest recorded longevity, as well as the relatively largest brain. The margay, on the other hand – a cat that lives in tropical forests – combines its slow development not only with a large brain but also exceptionally large eyes and short muzzles, two further infantile characteristics that combine with their bulbous heads to give this cat a kitten-like, 'cute' appearance. It is almost certainly no coincidence that, of all the other mammals on earth, the only group that approaches humans in their relative brain size is one that has gone through an impressive regression to an infantile stage – that is, dolphins and whales. From their lack of external limbs to their relatively undifferentiated skeleton – including poorly formed ribs and missing sternum – and their smooth and hairless skin and simple tooth buds, the general picture is that of an animal that remains suspended in an early developmental growth phase for the whole of its life. Through the simple process of radical infantising, and regression to an almost embryonic state, ancient four-legged herbivores that once grazed on open plains were thus transformed into the gargantuan tadpoles that now roam our oceans. Moreover, as Colin Groves points out, the brains of these animals are entirely in keeping with a hugely extended infantile growth period.

I am greatly enamoured of the idea that the humpback whale is an overgrown pig embryo with knobs on; and I find it revealing that in these neotenous [infantile] mammals the large brain is most plausibly viewed as not selected for as such, but is large by being, so to speak, swept along with the predominating trend.

Likewise, among non-human primates, our partially infantised relative, the bonobo, has a larger brain case than its more 'mature' relative, the chimpanzee. But what of our own species? Is there is any evidence that within the human species the degree of infantising is linked to brain size. For reasons given earlier, there is little doubt that the Mongoloids – a group that includes the Chinese and Japanese – have been more infantised than all other races. So do they also have larger brains? The answer appears to be yes. Despite this being a field that can become incredibly entangled with racist agendas, there is strong evidence that Mongoloid people – as well as having more rounded infantile heads – do indeed have slightly larger brains than other races. Of course, it is apt and propitious to follow this statement with a reminder that brain size is not in any way linked to intelligence.

The evolution of our big brain merely as a by-product of extended infantile growth periods also explains a rather curious finding with regard to the relative sizes of different parts of the human brain. It has long been assumed that humans are more intelligent than chimpanzees because we have a larger 'frontal cortex' – this being the part of the brain that is linked to intelligence. But recent research has shown that this is simply not true. It has now been found that the human frontal cortex occupies virtually exactly the same percentage of the total brain space as in chimpanzees, gorillas and orang-utans – around 38 per cent. Likewise, when considering the size of the human 'neocortex' – that part of the brain that covers most of the outside of the entire hemispheres, and which plays a crucial role in some higher forms of learning – we find that it is relatively no better developed than in other apes. Richard Passingham, now Professor of Cognitive Neuroscience at Oxford University, describes his investigation into the relative size of the human neocortex:

The volumes of the brain and neocortex are plotted for each species of primate, and a regression line is fitted to the data for the non-human primates. Reading from this line it is possible to work out the size of the neocortex that would be expected for a hypothetical primate with a brain matched in size with the human brain. The answer is striking: the point for man falls very close to this line. In other words our brain has no more neocortex than we would predict for such a primate.

Two supposedly critical parts of our brain are proportionately no larger than in any of the other great apes. This is of immense significance, since if there was strong evolutionary selection working directly on the brain – as would be expected if it was contributing significantly to our technological or social expertise – it would be expected that certain 'important' parts, such as the frontal cortex, should become more highly developed than others. But clearly this is not the case – our frontal cortex, and indeed our whole neocortex, is exactly the same size as expected. On the contrary, if our brain merely kept on growing for that much longer than in our closest relative, free from any evolutionary pressure and simply as a by-product of our infantised development, then it would be expected that components such as the frontal cortex should remain the same relative size – as, indeed, turns out to be the case.

Meanwhile, other parts of our brains have enlarged for apparently no reason at all. Referring to the particular part of our brain that interprets visual stimuli, Michael Gazzaniga, Director of the Center for Neuroscience at the University of California, Davis, commented, 'And yet, why is the primary visual cortex of a human several times the size of the monkey's? Abundant psychophysical measurements have demonstrated time and again that the monkey's visual system has virtually the same characteristics as the human system, indicating that a small visual system (cellwise) can do everything a large system can do.' From a theoretical point of view, a larger size might even hinder the brain's effectiveness. Evidence from computer science suggests that above a certain size, increases in complexity actually cause a reduction in processing power. As our ancestors' brains grew they could have become increasingly overloaded by the sheer bulk of information being processed.

In the light of all this information, it is now possible to revisit the question of the evolutionary mechanism by which our ancestors became infantised and, incidentally, acquired their massive brains. Having already established that sexual selection, by females, is by far the most likely driving force behind the infantising of our species – with both sexes possessing a variety of physical and behavioural features that support this view – does our inordinately large brain change the situation at all? The answer is no – it fits in perfectly with the rest of the 'package' of infantile characters that makes us who we are. The only difference between this and the first round of infantising – that which occurred many millions of years earlier when our earliest ancestors came out of the trees – was that now females were in control, they could select males not only directly, on the basis of their infantile behaviour, but also indirectly through their infantile appearance, since this provided a useful way of predicting the behaviour of a male. With females involved, it is hardly surprising that the second round of infantising of two million years ago was associated with a far greater change in the overall physique of our species.

With females selecting for both infantile looks and behaviour, the effect on brain size would have been considerable. Domestic animals, such as dogs, and Professor Belyaev's foxes, show how selecting for infantile behaviour alone can lead to an increasingly infantile appearance, even when the latter is totally irrelevant to the selection process. What is more, animals ranging from margays to tenrecs, whales and bonobos provide evidence that infantile species tend to possess larger brains. However, in the case of our own species, the effect of females selecting infantile male looks is likely to have had the greatest effect on the growth of the brain.

As David Perrett's experiments showed, women are most attracted to men with faces that combine masculine and baby-like features. And of all the features of the head that give it an infantile appearance, a large vaulted brain case is certainly among the most powerful. Even though having a big brain was of little use to males, it could – like the peacock's tail – provide females with important information about the male's infantile condition. In the case of the gaudy peacock's tail, this costly feature indicates that the male is sufficiently healthy and fit to be worth mating with. In the case of a large human head, this similarly costly feature indicates that the male is infantile and likely to form a strong pair-bond.

But females being attracted to infantile males was sexual selection like no other – it was more a case of super-sexual selection. Unlike other sexually selected features such as long fins in fish, colourful feathers in birds, or deep croaks in toads, big heads in humans not only stood as a reliable indicator of another quality – in this case, that a male was sufficiently infantile to form a strong pair-bond – it also made him resemble an infant himself and thus triggered powerful maternal instincts in the female. As males became increasingly infantile in their appearance, so the urge for females to look after and mother them would simultaneously have become stronger and stronger. More than fifty years ago, the great ethologist Konrad Lorenz proposed that juvenile features trigger 'innate releasing mechanisms' for affection and nurturing in humans, and suggested that this explains why we are drawn to animals that, purely by coincidence, share certain features with human infants – such as giant pandas with their huge and endearing black 'eyes'.

The compulsive attractiveness of excessively large heads is nowhere better illustrated than in America's most celebrated cartoon characters. Twenty years before Konrad Lorenz proposed the value of juvenile features in eliciting caring behaviour, Walt Disney and his artists had already begun to exploit this particular biological phenomenon by providing their two-dimensional stars with vast heads to ensure that the viewing public was irresistibly drawn towards them. In an essay entitled, 'A biological homage to Mickey Mouse', Stephen Jay Gould takes great delight in tracking the evolution of the world's cutest murine superstar. When Mickey first appeared in the late 1920s, he was mischievous and even displayed a streak of cruelty. But soon, to improve Mickey's mass appeal, his personality and appearance gradually began to change.

> [Mickey became] progressively more juvenile in appearance ... he assumed an ever more child-like appearance as the ratty character of *Steamboat Willie* became the cute and inoffensive host to a magic kingdom ... His head grew relatively larger and its features more youthful ... [and his] ears moved back increasing the distance between the nose and ears, giving him a rounded, rather than a sloping, forehead.

In our own distant ancestors, with large heads not only indicating a dependable partner, but also pulling on the maternal heartstrings of females, there would have been no stopping female preference for this character in their mates. What is more, it provided the perfect recipe for intense 'runaway' sexual selection. If infantile heads were attractive, then super-infantile heads were super-attractive – regardless of the cost. If females wanted large heads, then that is what the males had to have if they were going to attract a mate, and with daughters as well as sons inheriting their father's characteristics, so the whole species spiralled towards an ever more baby-like appearance – including, of course, having an inordinately large head. The brain, perhaps more than any other part of our body, stands as a highly visible and expensive monument to the power of female sexual selection for infantile partners.

THE DAWN OF CONSCIOUSNESS

For some, the journey of the human brain towards its present state is comparable to that of the Orient Express. Like the sybaritic train that started its journey in Paris and passed through numerous intermediate stages, including Munich, Vienna, Budapest and Bucharest, before finally reaching its final destination of Constantinople – our brain followed a gradual, albeit branching, journey towards its present condition. Personally, however, I do not subscribe to such an analogy. If such a parallel was to be sought, it would instead involve being a passenger on a train that progressed steadily along a gradual route until, one day, the passenger inadvertently steps off the train and finds him or herself in a rocket that is heading straight for the outer reaches of the Universe.

For hundreds of thousands of years, our ancestors' brains continued to grow unabated while they reaped few intellectual benefits from this ever expanding sexual attractant. Across the world, big-brained apemen successfully gathered, scavenged and hunted by means of simple and easily copied tools of the type that had been used for countless generations before them. These ancestors were especially successful not because of their advanced technologies or intellectual abilities, but rather because of their highly cooperative and cohesive social systems that were based around pair-bonding and sexual division of labour.

But, finally, something arose that had never existed before in the brain of any animal. It was something that transformed the already large brains of our ancestors into what they are today. While there is a good chance that this new type of brain circuitry arose somewhere in Africa, regardless of where it happened, it was so radical and advantageous that it spread like wildfire across the globe. Until this point, our ancestors had merely hit rock against rock to create a sharp edge. Suddenly, after millions of years of very little progress, they started to represent themselves and other animals in paintings on cave walls, make little statues, as well as bury their dead in a ritualistic way. From this point onwards, they were no longer simply clever apes, now they were indisputably human.

So what was the 'something' that happened to these ancestors' brains that gave rise to such radically different behaviour? For a start, the chances of a totally new structure emerging – along the lines of René Descartes' pineal coordinator, lying in the centre of the brain, which served as the gateway to consciousness – are, to put it bluntly, as likely as us sprouting wings. Despite many thousands of hours looking, scientists have yet to find a single part of our brain that does not also exist in other animals. It is therefore at our peril that we ignore the maxim, '*Nihil est in intellectu quod non primus fuerit in sensu*' – which translates as 'There is nothing in the mind that has not been previously in the senses.' To ignore this, and all its ramifications in terms of combining previous experiences, is to tread dangerously close to the ethereal. There can be, as philosopher Gilbert Ryle said, no 'ghost in the machine'.

On the contrary, it is well known that as brains get larger, so many components within these brains – including those that control touch, sight and movement – multiply. Big brains possess many more copies of identical units than small brains. The result is that big-brained animals can process more information at the same time than small-brained animals, and thereby come to more sophisticated decisions about the best course of action to take. Whereas a simple brain might see a dead animal and start eating it immediately, a big brain might combine this discovery with other information, such as trampled grass and fresh blood, conclude that there is a high probability that a predator is still in the near vicinity, and vary the behaviour accordingly. It is just like adding extra processors to a computer – the more processors there are, then the more tasks that can be undertaken at once, and the more complex the tasks that can be performed become. This is called 'parallel processing', and it explains why animals with relatively big brains, compared with the size of their body, are invariably more intelligent than animals with relatively small brains.

However, it seems that by the time a brain gets to the size of that of a chimpanzee – that is, about 500g – it has come extremely close to the zenith of its power. After this, as our ancestors ably demonstrated, even vast increases in the size of the brain and the number of copies of each processing unit make very little difference to mental skills. The only way to improve the computing power beyond this point is to evolve a new wiring configuration or – in the absence of a new hardware – a new software 'brain operating system'. What had occurred since our first ancestors left the trees were merely upgrades on the brain operating system that they had inherited from the common ancestor of themselves and chimpanzees. Indeed, chimpanzees today possess their own particular upgraded version of this ancestral operating system. However, for a serious increase in brain computing power, a radically new operating system needed to emerge, and it was only when this was launched that humans could be said to have evolved. What, then, made the new brain wiring configuration, or operating system, so very different from all the others that had proceeded it?

As I ran through the local park this morning with some friends, I spent embarrassingly little time listening to the conversation. Instead, my mind drifted and I was thinking about a variety of issues that needed to be resolved. My brain was, however, aware of what was going on around me, this being clear from the fact that any mention of my name caused an instant shift of my attention to my fellow runners. But for most the time my mind was weighted towards internal thoughts. In short, I was 'daydreaming' – staring at one thing, and thinking of something quite unrelated. A part of my brain was isolated from all that was going on around me and leaping from subject to subject, in total disregard of outside issues. Depending on which inputs were the strongest, my brain activity shifted more towards internal or external matters, however, the essential point is that it is able to analyse information coming from the two sources simultaneously.

The radical new wiring, or brain operating system, that transformed apemen into humans, was simply one that allowed a part of the brain to break free from external influence – to roam unconstrained from one memory to the next – while other parts of the brain analysed information coming in from the outside world. The brain, like a computer's processor, was 'partitioned'. It is the complex blend of these two distinct parts of the brain that creates subjectivity – a perception of the world that no one else will ever experience. It also provides us with 'consciousness', a sensation that results from our brain's unique ability to contrast, and thereby distinguish between, information arising from inside and outside our head. Without simultaneously experiencing an internal world – through daydreaming – and an external world, then there is no concept of 'inside' and 'outside', and thus no concept of 'me' and 'not me'. This ability of the brain to distinguish between internal and external stimuli may well contain a learned element, given the time that it takes for young children to be able to distinguish between their own thoughts and those of others.

This is not to say that other animals' brains can never become detached from external stimuli. They certainly can do this during sleep. Any dog owner will no doubt have observed their companion twitching away and even making muted barking sounds while fast asleep – something that clearly indicates that dog brains can relive events that have nothing at all to do with their immediate surroundings. The big difference, of course, is that dog brains contain a type of circuitry – or, alternatively, operating system – that allows them to process only one input at once. They can either interpret their surroundings, or dream, but not both at the same time. They cannot let their brains drift while interacting with the real world – in other words, daydream. It is inconceivable that a dog, for example, should suddenly start wagging its tail as it walks down the road, simply because it has just remembered something that happened the day before. We, on the other hand, because of our partitioned mind, are uniquely capable of both dreaming and reacting to our surroundings at the same time.

In volume 83 of the journal *Philosophical Review*, published in 1974, Thomas Nagel famously called his article on the nature of consciousness 'What is it like to be a bat?' Nagel's conclusion, however, which is equally famous, is that we will never know. According to Daniel Dennett, Director of the Center for Cognitive Studies at Tufts University in Massachusetts, 'we human beings don't have and could never acquire the wherewithal, the representational machinery, to represent to ourselves what it is like to be a bat.' But, accepting that we can never really know, how close can we get to imagining what life would be like without the human brain operating system? What would it be like to possess an alternative operating system that did not allow a part of our brain to drift free from all external stimuli? Perhaps the closest it is possible to come to this state of mind is to force ourselves to speak out loud the name of everything that catches your eye. Telephone, clock, calendar, book, pencil, window, television, penknife, video, CD, picture, key, paper... and so on. The result is that you are so busy with reality, so busy associating object with names, that your brain cannot mull over what happened yesterday or imagine going for a run this afternoon. You are incapable of actually 'thinking'. All your mind can do is jump from one association to the next, as dictated by what you are seeing, hearing or touching at that moment. However complex these associations may be, and however many variables you might be able to hold in your brain at once, your thoughts are rigidly shackled to what is happening around you. No wandering mind, no meandering sequence of memories that leads to thoughts of distant pleasures, no reminiscing, no fantasies. Just interpreting the here and now. This is most likely the closest that we can ever get to imagining what it like to be another animal.

The same logic applies to evolution of language. If all we can do is concentrate on one word at a time, it does not matter how many associations we form with any one word, we will never be able to understand a sentence. Understanding language, as with drawing conclusions about a room full of objects, requires us to be able to withdraw from what our eyes are seeing, or our ears are hearing, to make sense of patterns of words – to revisit past experiences without the distraction of the present. Daydreaming also gives us a sense of time. While other animals live only in the present, our brains can simultaneously experience, and therefore compare, present surroundings and daydreamed events of the past. Only by being able to juxtapose past and present experiences by these means, can any sense of time be gauged. All other animals live in a one-dimensional time – no past, present or future – just analysing the here and now.

If the existence of a partitioned brain makes us human, the relationship between the two parts – daydreaming and real time – very much defines our personal mental characteristics. Again imagining the situation as a single computer processor that has been split in two, the way the unit works is largely influenced by the amount that is given over to each part, as well as the level of communication between the two. Regarding the level of communication first, this clearly influences the amount that our daydreams interact with the present situation. Too little communication between the two parts and we have a limited ability to form abstract links between external stimuli and past memories. Too much communication, on the other hand, and it makes it difficult to distinguish between reality and fantasy. In its mild form, excessive incursions can appear as Freudian slips, as our daydreaming intrudes on reality, while in more severe cases it can produce schizophrenia – a condition that, of course, can be associated with highly abstract and creative thought.

As for the amount of the total processor that is given over to each part, it appears that our brain can devote more or less of its processing power to daydreaming or reality, according to the relative strengths of the inputs from each of these two sources. In other words, the partition that allocates processing power to internal or external matters can float. If our heads are full of highly active unresolved thoughts, then the internal activity predominates – and we daydream. In contrast, if we are being bombarded with powerful sensory information, then our internal activity is overwhelmed and we cannot 'hear ourselves think'. When we want to resolve a particular problem, we often want to be left to ourselves – we try to reduce the external stimuli so that internal matters can dominate our brain's activity.

To possess both abstract creativity and practical sanity can therefore be seen as something of a fine balance. If too much of our brain is devoted to daydreaming, or there is too much communication between the two parts, then we will find it difficult to concentrate on what is going on around us. Although it is something of a cliché, there is doubtless truth in the image of the creative genius as being somebody whose brain is so preoccupied with internal thoughts that they are largely oblivious to much of what is going on around them. For such a person, cascades of sounds and images coming from the recesses of their brain consume so much of their processing power that there is little left to deal with external matters. The image of the true creative genius as somebody who is sitting on the edge of insanity, as they maintain a tenuous and fragile link with the outside world, is one which many psychiatrists have recognised. Henry Maudsley, a psychiatrist who for much of his career petitioned against the 'insane' being allowed to have children, finally admitted an undeniable link between madness and brilliance.

To forbid the marriage of a person sprung from an insanely disposed family might be to deprive the world of a singular genius or talent, and so be an irreparable injury to the race of men ... If, then, one man of genius were produced at the cost of one thousand or fifty thousand insane persons, the result might be a compensation for the terrible cost.

In contrast, if too much of our brain is devoted to processing external sensory information, or there is too little interaction between the two parts, then we find it difficult to form abstract associations and to be imaginative. This condition can, however, result in a person being acutely aware of their surroundings. An extreme example of such a person is Solomon Shereshevski, the world's most celebrated 'memory master'. Shereshevski's talents were simply astonishing. He could recite hundreds of complex numbers after seeing them just once on a blackboard, and recall huge strings of numbers that he saw months or even years earlier. With his astonishing ability to concentrate on his surroundings, Shereshevski seems to have possessed a brain that was almost totally devoted to interpreting the outside world, and virtually incapable of daydreaming. Far from making him the life and soul of parties, and the source of countless stories and jokes, Shereshevski's mental state was actually more of a severe and sad affliction. He was almost incapable of thinking in the abstract. He became extremely muddled when a word had two meanings or an object had two names. His mind was restricted to straight-line associations, and could not wander to revisit a world of distantly related past experiences. In fact, he could hardly read. His doctor, Aleksandr Luria, wrote, 'Each word he read produced images that distracted him and blocked the meaning of a sentence.' He even found it difficult to relate to his wife and son who, according to Luria, 'he perceived as through a haze'. Far from being seen as highly advanced, Shereshevski's brain should more accurately be described as highly regressed. In being largely incapable of daydreaming Shereshevski was, in short, behaving more like a brilliant animal rather than a sophisticated human.

What Solomon Shereshevski, who is not noted for having an abnormal-sized head, as well as people who are exceptionally creative despite having small heads, show – is that human mental skills rely on an aspect of our brain that is unrelated to its size. Small brains that are appropriately partitioned, and good at daydreaming, are considerably more creative than big brains that are shackled to reality. For hundreds of thousands of years our ancestors' brains enlarged merely as a part of an infantile package – as an incidental and secondary effect of our species becoming more baby-like – with little tangible benefits in terms of mental skills. But then, possibly for reasons to do with our accidentally large brains possessing a large amount of redundancy in terms of replicated brain units, a new wiring configuration or operating system appeared. The out-of-control enlargement of our brain serendipitously led to a chance, unplanned, event. Consciousness, symbolic thought, art and creativity were the result of a simple fortuitous mutation – and once this had happened, there was no turning back.

For our ancestors to become truly human – to be conscious of their own existence and actions – they had to be able to daydream. Only with this talent could encounters with rocks and sticks, for example, tortuitously combine with daydreamed recollections of structures bound together by a vine, to create the new concept of a pickaxe. Such invention depends not on the simple re-enactment of learned behaviour – however sophisticated this might be – but rather on the chance collision of two previously unrelated pieces of information in the brain. While daydreaming makes this possible, consciousness gives us the illusion of control of our thoughts. The mind is not a dirigible thing, rushing from place to place like an aircraft under the control of its pilot – rather, it is vagile, like a weightless balloon that wanders around at the mercy of gusts coming from every direction. While one part of our brain is receiving and processing information from outside the body, the other part is merely responding to the pandemonium that is constantly going on in our heads. We do not choose what comes into our mind. Our thoughts – be they words, images or music – are merely those that happen to win the battle for dominance at that precise moment. As the German philosopher Karl Vogt proposed, 'The brain secretes thought as the stomach secretes gastric juice, the liver bile, and the kidneys urine.' Of course, the recipient of these secretions is the brain itself.

As for when such a transformation happened? The answer is astonishingly recently, and certainly within the last 1 per cent of our species' evolution. After hundreds of thousands of years of minor changes, around 50,000 years ago things began to change very rapidly for our ancestors. For the first time in the archaeological record, body decorations, bone and antler industries and long-distance exchange of raw materials started to appear.

In his book *The Madness of Adam and Eve*, neuroscientist David Norrobin approaches the subject from a slightly different perspective, and proposes that the emergence of schizophrenia – a disorder that literally means the possession of a 'split mind' – marked the earliest arrival of conciousness. While he considers that this may have taken place as early as 100,00 years ago, he similarly claims that it was a defining moment for the emergence of modern humans, 'the single and most important event in human history... the break-point between our large-brained, possibly pleasant but unimaginative ancestors, and the restless, creative unsettled creatures we so obviously are today.' Regardless of exactly when and where the spark of self-awareness first emerged, what we can be sure of is that by 50,000 years ago it was spreading like wildfire throughout the world. By now the revolutionary new brain operating system was fully operational. From this point onwards, there was no holding our species back, as our ancestors' new brain configuration enabled them to create increasingly sophisticated lifestyles. It was like taking the lid off a pressure cooker. Creativity exploded as we started to represent ourselves and our surroundings in art, make complex tools out of different materials, bury our dead in elaborate graves to prepare them for an after life, make boats, and so on. At the same time, by no coincidence, a massive population explosion occurred in our species.

For 99 per cent of our evolution since we separated from other apes, our massive brains had provided us with little more computing power than those of chimpanzees at a third of the size. Only after accidentally acquiring a modification that allowed its possessors to dream and be awake at the same time, did our sleeping giant of a brain suddenly become the biological super-computer that it is today. If this had not happened, our brains could be even bigger than they are, and yet we would be just as intellectually inept as chimpanzees.

chapter seven

the extreme child

With every step that our species has taken, more and more pressures have emerged to drive us in an increasingly infantile direction. To start with, it was the environmental benefits of living in large infantile non-aggressive and cohesive groups. Then sexual selection became involved as females selected infantile and dependent mates to provide for them and their offspring. And finally, with the advent of our conscious mind and our species' reliance on technology, yet another childlike quality – that of playfulness – became extremely valuable in adults. With pressure coming from all sides, the result is that we have become a highly sociable, pair-bonding and inventive species. It is an explosively successful package of infantile qualities that has propelled our species to unimaginable heights of success. But despite this inordinate success, we have ended up in a somewhat precarious position. We have become so infantile that we now risk failing to develop certain reproductive behaviours. This chapter is about the occasional loss of one of the most evolutionarily-important behaviours of all – that of heterosexual attraction.

THE GAY PUZZLE

In a corner of my office sits a small filing cabinet packed with the photographic fruits of my travels over the past twenty years or so. For the most part, it is filled with the eclectic variety of images that one would expect to find in a zoologist's photographic collection – a curious spider from Borneo that bears an uncanny resemblance to a bird dropping, an immense truffle-hunting pig from the Perigord region of France, and a female deep-sea angler fish that has tiny parasitic males fused to her body. Being catalogued alphabetically by country, there are thousands of photographs to wade through before the letter U is reached. But it is definitely worth it, especially for those who take pleasure in observing the more bizarre phenomena to be found in the animal kingdom. For here can be found extraordinary images of mature males mimicking females, males possessing pseudo-mammary glands, males displaying with flowers on their head, and even males courting each other with the aid of amazingly oversized false penises. More so than anywhere else in the world, here there is gaudy colour beyond belief, mimicry that puts bird-dropping spiders to shame, and adult playful creativity that is simply unknown in the rest of the animal kingdom. Labelled simply 'USA – New Orleans', this section of the filing cabinet contains a collection of photographs from the New Orleans gay parade during Mardi Gras.

I remember my trip to New Orleans in March 1996 with mixed

feelings. Despite my sheer delight as wave after wave of extravagant novelty passed me by, I also remember distinct pangs of inadequacy. Whichever tool I grabbed from my evolutionary tool kit, I was incapable of even beginning to dismantle the scene in front of me to better understand what was going on. This was my introduction to the Gordian knot of human homosexuality. Despite same-sex behaviour being observed in other species, such as in bonobos as mentioned earlier, no other animal exists in which a part of the population opts for exclusive same-sex relationships. For a variety of reasons, including those to do with dominance displays or the temporary absence of members of the opposite sex, males or females of species observed in the wild occasionally direct sexual behaviour towards members of the same sex, but this is very different from exclusively preferring to have sex, or form a pair-bond relationship, with members of the same sex. Interestingly, the only glimpses of same-sex preference in other species are in domesticated species that, of course, along with ourselves have been significantly infantised. And yet, homosexuality is a universal – albeit minority – characteristic of our species. It occurs as much in advanced civilisations as it does in primitive cultures and is found throughout the world. In the 'Terminal Essay' to his translation of the *Arabian Nights*, Sir Richard Burton provides endless examples of homosexual practices past and present, and claims that such practices have always been endemic over a vast area of the globe.

There have been a plethora of theories put forward to explain homosexuality in our species, from those who see it as a disease of the brain, to others who blame it on a disturbed upbringing. Sigmund Freud suggested that some boys become homosexual as a way of avoiding conflict with their dominant father. By pursuing another male, a homosexual son avoids competing with his father for the most available female – that is, his mother. Others have painted imaginary scenes, set hundreds of thousands of years ago, where homosexuality may have evolved to serve some function. Homosexuality, they suggest, could have evolved where a limited amount of food meant that some individuals, rather than having more children, devoted their energy to helping relatives bring up their offspring. Like worker bees in a beehive, instead of adding to the pressures of overpopulation, homosexuals would thereby have increased the chances of survival of their relatives' offspring. As for explanations for female homosexuality, psychologists fare little better. In fact, the situation is so bad that, as Geoffrey Miller wrote in his recent book *The Mating Mind*, 'No biologist has ever offered a credible theory explaining how exclusive homosexuality could evolve in a sexually reproducing species.'

None of the theories so far proposed seem at all convincing. The suggestion that homosexuality is an illness is simply out of the question – it is far too deeply embedded in our species' behaviour and far too old and widespread for this. There is also very little evidence that homosexuality is related to the type of relationship that a child has with its parents, making Freud's theory highly unlikely. Moreover, children who are brought up by homosexual parents are no more likely to become homosexual later in life than those raised by heterosexual parents. As for the 'population-control' arguments, or those that suggest that homosexuals are there to act as 'helpers' to look after young relatives, these are totally nonsensical given the fixed nature of homosexuality. If evolution ever did come up with a behaviour that dissuaded individuals from breeding under certain environmental conditions, it would have to be totally flexible and change along with the circumstances. Clearly, homosexuality does nothing of the sort, and it is absurd to think that homosexuals should suddenly become heterosexual when their standard of living, for example, improves. As it happens, there is already a biological mechanism that is quite capable of limiting reproduction when times are tough. If a woman's fat levels fall below about 15 per cent of her total body weight, or she is in some other way unusually stressed, then her whole menstrual cycle naturally grinds to a halt and she becomes incapable of conceiving.

Whatever else may be uncertain in life one thing is for sure: there is nothing more lethal in evolution than a behaviour that prevents, or even inhibits, an animal from reproducing itself or its genes – and that is exactly what exclusive homosexuality does. If a single gene ever arose, the sole purpose of which was to make somebody exclusively homosexual, it would simply never be passed on to the next generation. Given the amazingly high incidence and ubiquity of homosexuality in our species, the true explanation simply has to be far more fundamental as well as functional than any of the theories mentioned so far.

GROWING UP

Humans, like all mammals, go through a series of behavioural stages as they grow up. As an infant, when it is weak and fully dependent on its mother for its survival, an individual's behaviour is simplistically self-centred and devoted to ensuring the greatest possible attention for itself. At this earliest stage, an infant clings tenaciously to its mother and resists any attempt to be separated. Soon, as it grows stronger, the infant starts to develop friendships with other individuals. To begin with, it restricts its social forays to members of its immediate family, especially its brothers and sisters if it has any. Next comes relationships with those outside the family who are of a similar age. Initially, the sex of the playmate is unimportant; however, as they grow up they increasingly turn their attention towards forming close friendships with members of the same sex. This is a stage that can last for many years, during which males and females direct nearly all their attention and energy to the activities of single-sex groups or 'gangs'. This all-male or all-female stage is especially important among primates because of the intricate relationships that exist within their social groups.

Turning to our own species, it is possible put ages to these various stages. For the first six to twelve months of life, all our attention is directed entirely towards our mother – what thoughts we possess are solely aimed towards the comfort that we derive from being close to her. At around twelve months old, however, we are ready to proceed to the next stage of development as we start to show interest in the activities of others, especially others of roughly the same age as us. If sat alongside another twelve-month-old, we touch each other and pull at each other's hair and clothing. By eighteen months old, we are ready to embark on our first interactive relationships with others and we start playing with other infants in a semi-coordinated fashion, chasing each other and imitating the actions of those who surround us. During this early phase of our social lives, the sex of our playmate is largely unimportant, and even at four years old, we still think it perfectly reasonable for boys and girls to play with identical toys.

However, concepts of gender stereotypes soon develop, even in families that espouse gender equality. Indeed, the content of these stereotypes is remarkably similar in cultures around the world, despite considerable differences in social pressure. Psychologist William Damon demonstrated how children's attitude towards the 'gender appropriateness' of certain toys changes with age by telling four- to nine-year-olds a story about a little boy named George who liked to play with dolls. He recounted how George's parents told George that only little girls played with dolls and that it was wrong for little boys to do so. Damon then asked the children questions, such as 'Is there a rule that boys shouldn't play with dolls?' The answers he received were wildly different depending on the age of the children. The four-year-olds considered it fine for George to play with dolls – they said that there was no rule against it and that George should go ahead and play with dolls if that was what he wanted. In sharp contrast, the six-year-olds were shocked by any boy wanting to play with dolls and thought that it was wrong for George to play with dolls. By about age nine, the children had developed more sophisticated concepts of 'right' and 'wrong' and considered that while it was not 'wrong' for George to play with dolls, it was inappropriate for a boy to do so.

As we head towards five years old, the sex of our playmates becomes an increasingly important issue. By the time we reach five or six years old, roughly three-quarters of our friends are of the same sex, and by age seven to eight all our previous friendships with the opposite sex will have been vehemently severed, with all our close friends now belonging to the same sex. In one experiment, where children were given the choice of playing either with a member of the opposite sex or a member of the same sex but a different race, they almost exclusively chose to play with the child of a different race in preference to a member of the opposite sex. During the six to twelve-year-old childhood phase, the urge to form close bonds with members of the same sex is often so strong that the opposite sex is treated with total disgust and contempt. I remember well that in my own school girls were derogatively referred to as 'hags', and that the effect of brushing against one of these untouchables was to instantaneously contract what we cruelly referred to as 'hag pong'.

The separation of boys and girls into single-sex groups is a universal feature of our species. Boys play with boys and girls play with girls regardless of where in the world they live and which culture they belong to. But however cruel an antagonism between the sexes might seem, especially when it results in a child being pitilessly reviled and ridiculed by a gang of the opposite sex, there is a biological point to it. At the age at which it occurs, the membership of a single-sex gang can be extremely important for survival since no one individual is yet strong enough to be able to stand up for him or herself against a mature adult. On the other hand, as a member of a gang, a child has a far greater chance of getting what it wants than if he or she was alone. Groups of children are often frustratingly cocky and impervious to scolding by adults, whereas on their own they are far meeker animals. This cooperative strategy is undoubtedly of vital importance for many young inhabitants of inner cities all around the world.

Membership of all-boy or all-girl gangs is not only of short-term advantage; there may well be other, far longer term advantages. The zoologist Desmond Morris has suggested that a major function of juvenile gang formation is to reduce the chances of incest. His reasoning is that, by separating themselves from members of the opposite sex for a certain period, when the sexes do reunite they find each other novel and, as such, sexually exciting. Of course, the only members of the opposite sex who would not be novel, and therefore not sexually attractive, are the brothers and sisters to whom the juveniles have been exposed throughout the period of segregation in the course of normal day-to-day family activities. Another suggested function of this sex segregation is to work out social hierarchies before the onset of maturity when clashes can lead to severe injury or even death. Skills that are needed by one sex or the other can also be shared and perfected during this phase. Above all, it is important that all these things have happened by the time that sexual maturity arrives, after which attentions are turned towards forming relationships with members of the opposite sex. Once this stage of life is reached, relationships with members of the same sex will never be the same again.

During the adolescent stage of life, the arrival of sexual maturity is soon followed by a gradual dissolving of the all-male or all-female gangs, as members of these groups start to explore relationships with the opposite sex. This can be an especially stressful time of life, not least because of the recurring battles that take place between an individual's sex drive and his or her loyalties to gang members. Choices constantly have to be made between long-standing same-sex friendships and exciting new sexual relationships with members of the opposite sex. Of all the stages of primate social development, adolescence is that which most closely resembles the permanent behavioural condition of our species. At this stage, individuals still possess many valuable immature characteristics – such as playfulness, maternal dependency and reduced levels of aggression – while simultaneously being capable of reproduction. A typical primate enters the adolescent stage as a juvenile with hardly a care in the world other than who to play with next, only to emerge some years later as a mature adult obsessed by status and reproduction. It is this mature primate stage that humans must avoid at all costs.

THE EXTREME CHILD

So what has this all got to do with homosexuality, or more precisely with the infantising of our species? The answer is variation – the life blood of evolution. It is axiomatic that each of us is different, and that our children are different again. In every characteristic, however tiny – from the dimensions of each part of our face to our intelligence – there is variation. Without variation, there is no chance to be different – no chance for us, or our offspring, to excel – and no chance to adapt to new conditions. As long as we reproduce by means of sex, and reshuffle our genes at every generation, there will always be genetic variation within our species. What is more, given that the infantising of our species must be broadly under genetic control – because of the consistent and heritable changes that have occurred throughout our evolution – there will inevitably be variation in this feature too. Depending on the precise combination of genes that somebody inherits from their parents no two people will be infantised to exactly the same degree.

What happens, then, to those at the far end of the infantile spectrum – those who have had their infantile stages extended the furthest into their adult life? The answer, of course, is that such individuals will fail to develop as many mature behaviours as others who are less infantised. In just the same way that people vary in the exact extent to which they retain infantile physical features – such as hairlessness, a flat face, or a small nose – so there is also variation in the extent to which they retain infantile behaviours. And this is where the origin of homosexuality can be found. For those humans who are at the extreme infantile end of the spectrum, a small number will simply fail to develop beyond the stage of forming close relationships only with members of the same sex, the stage that reaches its peak during the transition between childhood and adolescence. Homosexuals, through their extreme infantising, fail to cross the Rubicon that separates childhood same-sex attachment from more mature heterosexual attraction. Such are the huge benefits of remaining infantile throughout adulthood – both in terms of our creativity and appeal to the opposite sex – that our species has come to lie precariously balanced on the edge of failing to make it beyond this same-sex stage of social development. It is, of course, a risk that is well worth taking. In return for all the advantages of perpetual youth, women face a mere 1 per cent chance, and men a mere 3 per cent chance, of failing to develop a heterosexual drive.

That homosexuality, like other aspects of our individual personality, has a genetic component is beyond doubt. It runs in families. Michael Bailey and Richard Pillard, from the Northwestern and Boston Universities, demonstrated this convincingly when they compared the incidence of homosexuality in male twins and in unrelated adopted brothers. Moreover, of the twins studied, half were identical and therefore shared 100 per cent of their genes, while the other half were merely fraternal and shared 50 per cent of their genes. In theory, if homosexuality has a genetic component, then the more genes that are shared, the more likely it is that two individuals have the same sexual orientation. The results were as clear-cut as they could possibly be. Identical twins are twice as likely to be both homosexual as fraternal twins that share half as many genes, and five times as likely both to be homosexual as adopted brothers who, despite being brought up in the same environment, share no genes at all. The difference in sexual orientation must therefore come from the genes.

This genetic situation with regard to homosexuality is perfectly understandable in terms of some people carrying more 'infantile genes' than others. It also explains how, if two considerably infantised, yet not quite homosexual, people should have children, then there is a good chance that one or more of their children will inherit the most infantile genes from each parent and thus become homosexual. Homosexuality is just one of a number of examples of genes that work well in certain combinations but which can be evolutionarily harmful if present in too concentrated a form. The 'sickle-cell anaemia' gene, for instance, is only effective in protecting people from malaria when it is in a diluted form. If a child inherits two copies of this gene, one from each of its parents, then it suddenly becomes lethal. In the case of homosexuality, while the situation is no doubt more complex because of the involvement of more than one gene, the principle is the same. Lisa Geller – a molecular geneticist from Harvard Medical School – has proposed a very similar genetic mechanism underlying homosexuality,

Homosexuality could [have been] created by an allele of a gene with a number of repeats hovering *barely* under the trip-wire that triggers it. Say your mother had sixty-two repeats …but in you, something happened … that bumped it up to sixty-three. And that was the line. And that's why you're gay. But your brother has only sixty-two, so he's straight. So heterosexuals are carrying this allele around, passing it on, and not expressing it because they're hovering right under the line without knowing it.

The 'line', of course, that Geller is talking about, is the particular degree of infantising beyond which people fail to mature past the single-sex attachment stage.

Homosexuals are, of course, merely at the end of a continuous spectrum, in terms of the amount that different individuals within a population are infantised, and a far more common situation is that this early same-sex stage extends only slightly beyond the onset of a sex drive – before, finally, being replaced by heterosexual attraction. Evidence for this comes from statistics such of those of Alfred Kinsey and his colleagues which show that as many as 30 per cent of adolescent boys have homosexual experiences that lead to orgasm. Similarly, another study of nearly 35,000 youths in Minnesota, found that as many as 10 per cent said that they were 'unsure' of their sexual orientation – a figure that is clearly far higher than the 3 per cent incidence of homosexuality later in life. Given these figures, it is hardly surprising that many happily married men reveal that they had homosexual experiences when they were growing up. One report found that 40 per cent of men who had had sex with another man went on to add that this took place before they were eighteen years old and that it had not happened since. The unavoidable conclusion is that it is remarkably common for members of our species to enter adolescence in a state of sexual ambivalence. A recent survey of the sexual lifestyles of the British reported that:

These findings at least support the view that, for some, sexual development is characterized by a labile stage of orientation preceding a later stage of greater stability. A form of bisexuality prevalent in early adulthood may represent a transitional phase in which preferences are tested through experimentation with different lifestyles and relationships. Kinsey et al (1948) interpreted their data as showing that homosexual activity peaks in the early teens and declines slowly with increasing age.

So, while many humans will experience a transitory phase of homo-sexuality during adolescence, others at the furthest end of this infan-tile extreme will remain at this stage for life. Far from developing a 'new' behaviour pattern, homosexuals merely remain at a slightly earlier stage of social behaviour than others. Since infantising has been at the heart of our species' evolution, it could certainly be claimed that homosexuals simply possess a little more of the 'essence of humanity' than the rest of society.

What, then, is going on inside the bodies, and especially the brains, of homosexuals? What has evolution done to our species to prevent some individuals from developing to the stage of being attracted to members of the opposite sex? The available evidence suggests that our brains have become progressively less sensitive to developmental hormones such as testosterone. Rather than selecting men who had low levels of testosterone, for example – and therefore made more faithful and infantile partners – generations of females may actually have been selecting men whose brains were less sensitive to whatever levels of testosterone might have been present in their bodies. The brains of such men were essentially numbed to the hormones that previously drove them inexorably towards maturity. If so, this could mean that homosexuals' brains are less responsive to developmental hormones than heterosexuals' brains, which is why their develop-ment takes a slower and less complete route. Similarly, it could mean that the genes that control our species' infantising may be operating by partially switching off our brain's sensitivity to certain hormones.

Such a conjecture certainly tallies with the various studies that have failed to find significant differences in the levels of sex hormones in the blood of homosexual and heterosexual men. What it also allows for is a certain amount of influence from the environment. Even if a particular person has a low sensitivity to certain hormones, if their body produces particularly high levels of these hormones later in life for whatever reason, then this could cancel out the brain's low sensitivity in terms of the rate of development. Conversely, if a person has a higher than average brain sensitivity, very low levels of hormones later in life could drastically reduce their rate of development. What is more, there is considerable evidence that our brain's sensitivity to those hormones controlling our sexual development is largely set well before birth. Testosterone, for example, is known to have a major effect on the organisation and development of a number of regions in the rat brain, and so levels of this hormone during critical periods of development could have long-term effects on behaviour later in life. Simon Le Vay, author of *The Sexual Brain*, considers that there is much to commend a model based on differing sensitivity to vital hormones, in which

> there are intrinsic, genetically determined differences in the brain's hormone receptors or in the other molecular machinery that is interposed between circulating hormones and their actions on brain development. First, this would provide a mechanism that involves hormone-induced brain differentiation but which does not require there to be differences in the actual levels of hormones. Second, since there are several different receptors involved (including the androgen receptor, the estrogen receptor, and at least two 'estrogen-related' receptors), there is opportunity for selective effects on different brain systems.

Le Vay goes to some length to point out that homosexuals are far from 'brain-reversed', and that nearly all their behaviour is entirely typical of heterosexual people of the same sex. And clearly this should be so, since they follow exactly the same developmental route as any other members of their sex, up to the point at which the rest of the population develops an attraction towards members of the opposite sex.

A low sensitivity of certain parts of the brain could have two major effects on the development of the body. Firstly, during the early stages of life within the womb, it could effect the structural development and growth of different parts of the brain. Indeed, Le Vay has identified at least one part of the brain, in the hypothalamus called INAH 3, that is different in size between homosexual and heterosexual men. Having a small number of receptors for certain hormones could also mean that homosexuals' brains are more easily disrupted by fluctuations of hormones that would, in other less infantised members of the population, have less of an effect on the development of the structure of the brain. The result would be to make it more likely that homosexuals show less well-defined gender-typical behaviour. In itself, however, such an effect has no reason a priori to lead to homosexual behaviour later in life. Indeed, as Simon Le Vay points out, homosexual behaviour is to a large extent totally independent of whether a person's behaviour is gender-typical or not. Not only do most homosexuals possess gender-typical behaviour that is identical to that of heterosexuals, it is sometimes even exaggerated. For example, it is well reported that many male homosexuals show excessively male behaviour in regard to their tendency to seek a prodigious number of sexual partners. Typical female behaviour is to have fewer, not more, sexual partners.

On the other hand, the second effect of having a low brain sensitivity to certain hormones is undoubtedly linked to the emergence of homosexual behaviour, since it ultimately causes an individual to remain in the single-sex stage for life. As mentioned earlier in relation to the evolution of our species' infantile bodies, one of the most noticeable effects of infantising is the extension of the early stages of development. Humans, for instance, have far longer periods of infantile growth and behaviour than our non-infantised relatives, the chimpanzees. In possessing brains that are less responsive to certain hormones, homo-sexuals would therefore follow more gradual paths to full development – with each of their developmental stages, especially those of childhood, taking longer to give way to the next. At an early stage of life, this would combine with any possible effects that infantising has had on the brain, to make the child feel 'different'. This would explain why, for example, it is common for homosexuals to claim that they felt different from other children at a very early age, and why they are frequently teased at school for failing to conform to age-typical behaviours.

Regardless of whether their behaviour will ultimately be gender-typical or not, a strong reason for the nonconformity of homosexuals is that they are developmentally immature compared with others of their age, since they take longer to progress from one social stage to the next. One result, for example, of taking longer to develop a desire to associate only with members of the same sex – something that normally occurs between the ages of three and five – is that a child misses getting on to the merry-go-round that carries them so rapidly away from the previous gender-irrelevant stage of development. As their more precocious peers are accelerating away at breakneck speed through mutual reinforcement and gender-exaggeration, those at the more infantile end of the human spectrum languish innocently in the stage of being content mixing with members of both sexes. Numerous studies have reported that future homosexual boys are more likely to have been loners and to have been rejected by other boys. As for join-ing in with members of the opposite sex, this too is extremely difficult as they too have formed tight single-sex groups.

Caught in limbo, those who fail to join single-sex groups at precisely the right time are launched into childhood torn between the identities of both sexes. To make matters worse, because of the self-reinforcing nature of these gender merry-go-rounds, they become increasingly difficult to jump on to as time goes by – regard-less of how strong the desire to do so becomes later in life. Rejected by members of their own sex, it is inevitable that many fail to develop some of the sex-specific behaviours that arise through the constant competition that occurs within single-sex groups, explaining why gays and lesbians often reject stereotypical 'masculine' and 'feminine' lifestyles. The situation is summed up by one survey which concluded that 'few [male] homosexual respondents said that they were "femi-nine" while they were growing up; most described themselves as neither especially masculine nor feminine'.

The rest, as they say, is history. At the end of childhood, when other individuals leave the single-sex stage to explore heterosexual relationships, homosexuals remain one step behind as they continue to look for same-sex friendships. Depending on the degree of infan-tising, and the amount that their development has been slowed down, some people will linger for a while within this single-sex stage, while others will remain at this stage for life.

THE PLAYFUL MIND

Evidence for the youthful qualities of homosexuals has been staring us in the face for years. From Plato to Sappho, Leonardo da Vinci,

Michelangelo, Oscar Wilde, Noël Coward, Virginia Woolf, Gertrude Stein, Francis Bacon, T.E. Lawrence, Amy Lowell, Benjamin Britten and Rudolf Nureyev, there is an extraordinary association between youthfulness, childlike imagination, rebelliousness and homosexuality. Of course, it is entirely to be expected that homosexuals, having had their social development curtailed slightly earlier than heterosexuals, should remain especially playful, creative and imaginative. It should come as no surprise, therefore, to find that a disproportionate number of homosexuals work in the performing and graphic arts, and that technical hobbies are rare among homosexuals whereas cultural and artistic hobbies are far more prevalent.

It is certainly nothing new that homosexuality is more widespread among the more creative circles of society; in fact, this is so much the case that it is something that society has largely taken for granted. Ever since ancient Greek times, homosexuality has been inescapably associated with the more playful and creative professions – the arts, philosophy and science. Plato, for example, who lived around 400 BC and is regarded as the initiator of Western philosophy, wrote at length of the delights of male love in his famous and highly influential book *Symposium*. The word 'lesbian' itself comes from a female literary society that was started by the poetess Sappho around 612 BC on the island of Lesbos. Indeed, today the widespread adoption of the word 'gay' strongly suggests that homosexuality is linked to a somewhat more playful character than is the norm within the rest of society. A recent gay and lesbian programme called *Gaytime TV*, for example, was billed as 'the playful magazine primarily for lesbians and gay men'.

While the 'macho' mature males occupy more aggressive and status-related positions within society, those with more playful and creative characteristics tend to be attracted to the arts. In a speech at a gay rights rally, American comedienne Robin Tyler remarked that 'if Michelangelo had been straight, the Sistine Chapel would have been wallpapered'. Likewise, Vladimir Horowitz observed, 'There are three types of pianists: Jewish pianists, homosexual pianists, and bad pianists.' Of course, this is not to say that extremely infantile individuals have the monopoly over creative pursuits, but merely that the most creative minds should belong to those at the more immature end of the behavioural spectrum, and that this category will inevitably include a higher than average percentage of homosexuals.

Behavioural flexibility is one of the most striking characteristic of youth. With age and maturity comes an inflexibility that makes learning new behaviour patterns increasingly difficult, as epitomised by the saying 'You can't teach an old dog new tricks'. On the other hand, young animals are highly receptive to new behaviours and can absorb a huge variety of different behaviour patterns. In nature, of course, such flexibility is vital to adapt a young animal to its environment by enabling it to copy the behaviour of others, including the social nuances that are required when living in large groups. Mimicry is essential to the learning of virtually all complex behaviours, from effective feeding techniques to appropriate facial expressions. And so to the role playing, acting and mimicry talents of the most youthful members of our species. Role playing is a characteristic that is inescapably linked to homosexuality. Indeed, to many people, homosexuality is almost defined by the colourful, eccentric and often gaudy mimicry that is displayed at gay rallies. Although it is all too easy to succumb to popular stereotypes – including that of gay men dressed in leather and 'dykes on bikes' – it is an irrefutable fact that gay and lesbian rallies characteristically have far more of a 'dressing-up' and theatrical element than gatherings of just about any other group of people. In contrast with the earnest drabness of gatherings of politicians, farmers, health workers or civil servants, gay and lesbian events so often provide amazing spectacles of dressing up, role playing and rebellious behaviour that is as unrivalled and exuberant as it is childlike and adolescent. But role playing goes far deeper than this, as described by the academic and author Arlo Karlen in his exhaustive study of homosexual lifestyles:

> Role-playing, in fact, is one of the most important characteristics of the gay world. When a man acts the role of a petulant, flamboyant woman or a swaggering tough or an appeasing little boy, he may create stage settings to support the parts. One feels this in the self-consciously exotic atmosphere of many gay restaurants; in the melodramatic décor of many gay apartments; in gay clothes that are a bit like costumes.

The world of homosexuals who take part in overt mimicry bears astonishing resemblance to that of children who emulate and exaggerate each characteristic of their chosen role model.

> Anyone who spends any time with the Hell's Angels knows the difference between outlaw motorcyclists and homosexual leather cults. At any bar full of Hell's Angels there will be a row of sleek bikes lined up on the curb outside. At a leather bar there are surrealistic renderings of motorcycles on the wall and perhaps, but not always, one or two huge, accessory-laden Harleys parked outside – complete with windshields, radios and red plastic saddlebags. The difference is as basic as between a professional football player and a rabid fan. One is a performer

in a harsh, unique corner of reality; the other is a cultist, a passive worshiper, and occasionally a sloppy emulator of a style that fascinates him because it is so hopelessly remote from the reality he wakes up to every morning.

It is with good reason, therefore, that those at the most infantised humans – with homosexuals at the extreme – should excel and be overrepresented in the performing arts such as the theatre, television and ballet. Such professions, after all, have at their heart a childlike ability to mimic the details of others' behaviour.

The link between homosexuality and a protracted period of child-like curiosity is also illustrated by the academic achievements of homosexuals. It is entirely to be expected that as well as remaining behaviourally flexible and playful for longer than less infantised humans, homosexuals should also remain intellectually receptive and flexible for a greater percentage of their lives. Freud noted that homosexuals are 'distinguished by specially high intellectual develop-ment and ethical culture'. In fact, the results of surveys into the academic achievements of homosexuals are not simply impressive, they are staggering. In contrast with an average 'general male popu-lation' figure of roughly 8 per cent for college education, the percentage of homosexuals who achieved this level of education ranged from 53 to 91 per cent in different studies. At the Harvard Business School, gay people were nearly twice as likely to be ranked in the top 5 per cent of their class, and many businesses, such as Ben and Jerry's Ice Cream and Lotus, now offer domestic-partner bene-fits specifically to attract the top gay graduates. At even higher educa-tional levels, it is the same story. One study found that 13 per cent of homosexual men had Masters degrees, compared with a national average of 2.5 per cent, while 8 per cent had PhDs compared with a national average of 0.5 per cent. For lesbians, the trend is equally as striking. According to a survey carried out in 1992, the percentage of lesbian college students in America was 4 per cent – that is, eight times higher than among high-school graduates. Even the pejorative term 'pansy' which is sometimes given to male homosexuals, can actually be interpreted as something of a compliment, since it appears to have its origins in the French verb *penser*, 'to think'. One can imagine those wishing to denigrate homosexuals referring to them as the 'thinkers', as distinct from those who, in their opinion, contribute more by being employed in manufacturing industries. As it happens, it is quite extraordinary how few homosexuals do work in manual industries. But more of this in a moment.

Homosexuals' youthful qualities – especially their extended cooperative and social qualities – are nowhere better illustrated than in their job preferences. I have to say that I personally do not subscribe to the theory that gays and lesbians gravitate to certain professions simply because these uniquely provide a 'climate of tolerance'. Such a glib explanation simply begs the question of why gays and lesbians were attracted to these professions in the first place to establish such a tolerant atmosphere. This view is supported by the fact that direct job discrimination has been shown to affect a relatively small percentage of gays and lesbians. In an investigation into discrimination against male homosexuals – a study that had 1,057 respondents – Martin Weinberg and Colin Williams discovered that only 16 per cent of gay men felt that their homosexuality had an adverse effect on their jobs, while 5 per cent lost their job on more than one occasion when their sexual orientation became known. Clearly, there remains the possibility that discrimination prevents a considerable proportion of homosexuals being offered jobs in the first place; however, it is generally considered that young homosexuals, at least, are able to successfully negotiate the initial hiring process. No, rather than being channelled and forced into working in certain professions, a far more tenable explanation is that homosexuals' persistent youthful qualities, as well as creative and gregarious talents, attract them to certain professions above others.

Every survey that has investigated the job preferences of gay men has concluded that they very rarely work in manual industries. In 1961, it was found that while 55 per cent of the total male American workforce worked in manual industries, among gay men the proportion who performed this type of work was a mere 7.5 per cent. Since then, a number of other surveys have found exactly the same phenomenon. Instead of working in manual industry, gay men are hugely over-represented in white-collar jobs – including management, teaching and the arts. Similarly, gay men are vastly over-represented in service professions, including – clichés and stereotypes aside – working as male nurses, waiters and hairdressers. One report on the chosen professions of self-employed gays concluded: 'Typical lines of work for self-employed gays are those of hairdressing, barbering, pet care, floristry, and the ownership of a variety of other retail establishments.' In the few studies that have looked at the careers of lesbians, it has also been found that there is a strong bias towards those professions that involve working closely with people as well as an exceptional degree of social acuity and sensitivity. A recent survey found that one in seven lesbians is a social worker, while one in nine teaches.

So why the enormous bias of homosexuals away from manual industries, and why such a strong bias towards management, the arts, teaching, caring and service industries? One author has suggested that qualities such as 'toughness, strength, and physical aggressive-ness seem to have been rejected by many gays as not particularly desirable characteristics', while another commented that gays are often 'striving for gentility' and 'niceness'. Of course, both these authors are right inasmuch as men who remain at the most youthful end of the spectrum would be expected to avoid mature 'macho' professions in which bravado, aggression and physical strength – in other words, mature primate behaviours – are highly valued qualities. This typical lack of aggressive tendencies in homosexuals was amus-ingly described by Dick Leitsch in a letter to the editor of *Playboy*:

We homosexuals have enough prejudice against us without having to be blamed for including these weirdos in our group … We all know that the Chicago police is composed only of virile, heterosexual men – after all, what queen would strike or tear-gas women and nice-looking young men? As a matter of fact, from the havoc wreaked by the known heterosexuals around us, isn't it about time that the National Institutes for Mental Health granted some funds to discover whether murder and mayhem and violence is part of the heterosexual personality, or just a symptom of disturbed heterosexuality?

But more than this, maturity in men is typically associated with a far greater resistance to being touched, an increase in personal space and a general rejection of all signs of dependence. Ease with touching, as well as being touched, is a feature that is unquestionably associated with earlier stages of development – it is a vital part of the comfort-ing behaviour that is so important during an individual's early years. A desire for close contact is not only vital for a normal relationship between an infant and its mother, it also encourages the young animal to find protection by being close to adults. Even when adults are not present, comfort and safety in numbers can be gained by close proximity with other young individuals. The trend with age towards an increasing unease with close contact is frequently evident during the transition from childhood to adolescence, especially between members of the same sex. While it is quite normal, for example, to see five-year-old boys holding hands, the chances of teenage boys feeling comfortable with this type of close contact are extraordinarily slim. Even though later in life we re-establish an intimate sensual relationship with our sexual partner, most men still avoid showing signs of dependency, especially being touched, in public. Gay men, on the other hand, because they retain an excess of youthful qualities, stop short of developing the same reluctance and resistance to touch and be touched, thus explaining why so many are attracted to caring professions that involve social as well as physical contact, such as teaching and nursing. Clearly, reduced levels of mature assertiveness, dominance and aggression means that male homosexuals are far more comfortable than male heterosexuals in professions that involve dealing directly with other people. Indeed, research has shown that homosexuals are generally considered by others to excel in their handling of social relationships.

There is one more possible consequence of homosexuals being at the most youthful extreme of our species, and this concerns their body build. Since full maturity brings with it a bulking up of the body's muscles and bones – especially in men as they proceed through their teenage years – it can be predicted that if any members of our species fail to develop these mature features completely, then it should be homosexuals. And this indeed may well be the case.

Several investigators have discovered that homosexual men are significantly less well developed, in terms of their musculature and bone
development, than the rest of the population. Psychiatrist Ray Evans,
for example, found that homosexual men were, on average, 8 per
cent lighter than expected, and that they also had less muscle and
bone development, including narrower shoulders. They are what
anthropologists would describe as more 'gracile' – just as each 'new'
species of human ancestor was typically lighter bodied and more
gracile than the previous one. Most impressively, Evans discovered –
using a hand-grip device called a 'dynamometer' – that the homosexuals in his study group were 18 per cent weaker than expected. As
for homosexual women, investigators have found no consistent
differences between their bodies and those of the rest of the population – which is hardly surprising since women's muscles change far
less than men's during adolescence.

GOD THE FATHER

To a species of insecure individuals held in various stages of youth
and insecurity, the extended 'family' of the Church, with its guiding
Father figure at the head, is a necessary provider of comfort and security. From the local vicar being referred to as 'Father', to 'God the
Father', every aspect of religion has been designed to satisfy our
species' extraordinary need for parental protection and a craving for
guidance regardless of age. Hand in hand with our childlike imagination and inventiveness, has come the burden of lifelong insecurity
and a need to be looked after. More dangerously, it has also led to a
desire to hand over responsibility for our actions to a higher ranking
body, be it real or imaginary. Every culture in the world practises

some form of religion, and naturally those at the more immature end of the human spectrum would be expected to have the greatest need for such an institution.

It should come as no surprise therefore that there is a strong link between the Church and homosexuality, since both are directly related to our species' more infantile qualities. To the most infantile among us, in particular homosexuals, religious institutions offer the perfect social environment – under the watchful eye of a benevolent spiritual parent figure, they can while away their days surrounded by members only of the same sex. Childhood bliss. In fact, it seems that homosexuality and religion have been inseparable for much of Christianity's history. The German invaders of Roman Europe, for instance, as well as the Muslims who came into contact with Christians in medieval Spain and the Protestants from the time of the Reformation, all insisted that Catholic priests tended to be gay. Many other observers, with no obvious axes to grind, also noted the prevalence of homosexuality in the Church, including St Peter Damian in the eleventh century who wrote a long treatise entitled 'The Book of Gomorrha' about homosexuality in the priesthood.

From time to time there have, of course, been brutal attempts to rid the Church of homosexuality. Aside from their various other activities, members of the Inquisition burned vast numbers of homosexuals alive to 'save their souls from eternal damnation'. But despite the numerous cleansing sprees that have occurred throughout history, homosexuality has persistently resurfaced within the Church's ranks. As one author described the situation, 'With the arrival of the eighteenth century, it became evident that, in spite of having undertaken the most horrific holocaust the world had thus far known, the Church had lost its battle against pre-Christian beliefs and practices as well as against persons engaging in gender-variant and homoerotic behaviours.'

Recent studies have reported staggering levels of homosexuality in the Church. Among Catholic clergy in the USA, it has consistently been indicated that between 20 and 50 per cent are gay, and even when the extremes are discarded, a figure of between 20 and 30 per cent still seems to be most accurate. This is, of course, an astonishingly high figure when one considers the 3 per cent incidence of homosexuality among males in the general population. On a broader scale, it is remarkable how many religions throughout the world have incorporated homosexual acts into their ceremonies – from northern Moroccans who consider that a boy cannot accurately learn the Koran until his teacher has had anal sex with him, to Papua New Guineans who have rituals that involve boys drinking the semen of their elders. In many North American tribes it is the berdache – men who dress and behave like women – who enjoy the highest religious position. It is they who are supposed to excel in all shamanistic practices, even that of ventriloquism which is barred to normal women. A similar situation exists in South-East Asia, where in Borneo the highest grade of shaman is the *manang bali* – literally meaning 'changed shaman' – a man who has changed his sex to become female. Because *manang bali* have the highest rank in society, and are highly valued as doctors, they accumulate a considerable wealth; in turn, this leads them to be highly sought after by other men to be their 'wives'. However, to be the 'husband' of a *manang bali* can be something of a miserable life, since they are degraded to a very humble status in the household. It is particularly significant that within many religions around the world, it is homosexuals who hold the highest rank. Because of their greater retention of youthful characters, it is hardly surprising that homosexuals should be revered for their extraordinary 'visions' and 'insights' – or, put another way, for their childlike imagination, creative talents and play-acting abilities.

EXCEEDINGLY HUMAN

After hundreds of thousands of years of becoming increasingly infantile, it seems that our species has just about hit the end of the road. In essence, we have just about reached the point at which we have lost so many of our mature qualities that our ability to reproduce has become severely threatened. Primate babies might be incredibly sociable, inventive, creative and flexible in their behaviour, but they also have yet to develop the urge to reproduce. Over countless generations, our species has had so many of its immature behaviours extended into adulthood that we now run the risk of remaining in such a playful and immature phase that we fail to develop properly our urge to breed. It is ironic that female selection, a force that has been at the heart of the infantising of our species, has inadvertently resulted in some of the most attractive men of all becoming homosexual. Those males who possess the greatest number of childlike qualities – and who would potentially make the most attentive, dependent and caring partners – incidentally fail to find females sexually attractive.

Homosexuality is just one of the inevitable consequences of our species having regressed into a permanent state of infancy. Far from suffering from a disease or medical problem, homosexuals are simply the carriers of the 'youthful essence' of humanity in a more concentrated form than the rest of society. If infantising can be said to be at the heart of human evolution, homosexuals have been 'humanised' slightly more than the rest of the population. Of course, from an evolutionary perspective, occasional homosexuality is a small price to pay for sailing so close to the wind with regard to keeping the whole of our species' behaviour as immature as possible. For the most part, evolution manages to 'get it right' in producing playful and inventive, yet reproductive, individuals. Yet, for the 1 per cent of women or 3 per cent of men who just fail to cross the threshold to opposite-sex attraction, their gain is arguably far greater than their loss. As inheritors of our species' unique qualities in their most concentrated form, they have the potential to be among the most creative, innovative, and therefore most valuable members of society. When questioned about what special talents he possessed, the openly gay pop star George Michael replied, 'You don't understand. It's not that there's something extra that makes a superstar. It's that there's something missing.' The missing 'something' that gives us restless, insecure superstars, is maturity. Moreover, so long as homosexuality exists, we – as a species – are safe. It is only if it disappears that we are in trouble, for then we will inevitably be making our way back down the very dangerous road to maturity.

chapter eight

it's a
weird world

Homosexuality is just one of a variety of consequences of our species' infantile nature. Depending on which infantile qualities are retained to the greatest degree, we can express this defining human quality in a myriad of different ways, ranging from the endearing – such as the besotted lover – to the socially opprobrious, in the case of paedophilia and incest. It is to the side effects of our species' slide into nature's cradle that we turn now.

VENUS IN FURS

Although Leopold von Sacher-Masoch was more preoccupied by submissive/dominance games and fetishism, as opposed to pain-oriented sadomasochism, he undeniably deserved the accolade of having one of the most peculiar, and at first sight contradictory, human pleasures named after him. To Sacher-Masoch, sexual frisson was inexorably linked to submission and masochistic delight. To acquaintance Emilie Mataja, for instance, he beseeched, 'You would give me very great pleasure if you'd send me a good photograph of yourself. I would be even more delighted if you allowed me to visit you in Vienna and if you were willing – in furs, of course – to whip me.'

At the very mention of the word 'masochism', most people will vehemently deny that this behaviour has anything whatsoever to do with their love life. Rather than being erotic, masochism is perceived as being anathema and alien. They see no similarity at all between their relationships – that are built around love, gentle care and attention – and those of masochists who are obsessed with pain, humiliation and suffering. But they are wrong. In fact, the vast majority of us are on the edge of this particular sexual variation, since beating, bondage and apparently degrading physical and mental humiliation are but a stone's throw away from 'innocent' cuddling.

To Freud, masochism was 'the most frequent and the most significant of all perversions'. As a sexual predilection, masochism is far removed from sadism – a behaviour in which sexual excitement is derived from inflicting pain on another. Masochism, however, is by far the more common of these two sexual variants, with sadism most frequently occurring merely as a necessary adjunct that allows the re-enactment of a masochistic fantasy. Xavier Hollander, a media spokesperson for prostitutes, claims that roughly 90 per cent of the clients who purchase sadomasochistic services request the submissive role, while it has also been reported that among prostitutes who cater to rich and powerful clients in Washington DC, requests to be beaten outnumber those to inflict beatings by about eight to one. Significantly, masochism is also an activity that is practised more by men than by women, and male masochists request more intense pain and degradation than female masochists.

Despite the enormous efforts of sexologists, psychologists and sociologists, there have been few convincing explanations for the strong link that frequently exists between physical and mental pain and sexual gratification. The problem, however, is in the definition. When alternatively described in terms of a desire to 'experience helplessness', or 'be subjected to humiliation and submission', masochism might start to sound vaguely familiar. This should become unquestionably familiar if masochism is envisaged merely as an extreme version of a strong desire to identify somebody who holds far greater power than another for the purposes of protection. When viewed from this perspective, it becomes clear that an attraction towards domination, and even physical and mental torture, can be the direct result of a desire to be comforted and a need to satisfy immature feelings of insecurity. Although tender love and attention is clearly the most obvious way that a person can be provided with comfort, it can also be derived from evidence that somebody who is close to them is powerful enough to protect them. The situation is no different from that of small childen who are comforted not only from the love that their parents show them but also from the knowledge that their parents are strong enough to protect them from potential dangers if necessary. Despite the threat of chastisement if we should irritate or enrage our parents, their presence and strength remains our most powerful source of comfort. Indeed, it is often only at times when our parents are infuriated with us during childhood, that we become fully aware of their power, strength and dominance.

The origin of masochism clearly lies in our species' infantile insecurity. Since maximum sexual gratification is inextricably linked to feelings of comfort and safety, proof of a sexual partner's protective powers in the form of dominance displays can quite understandably bring about enhanced levels of sexual enjoyment. Once again, this results from the need for sexual partners to act as a pseudo-parent as well as provider of sexual gratification. As for the accoutrements and paraphernalia associated with this sexual variation, these too are understandable if we think in terms of the masochist as someone who retains a particularly strong need for parental protection throughout adulthood. Whereas children are able to associate power with the greater size and strength of adults, this is not possible with adults, hence the involvement of authoritarian clothing such as black leather, and instruments of aggression such as whips and handcuffs. In what might be considered a more extreme form of masochism, a minority of people obtain sexual gratification at an unmistakably infantile level simply by recreating a situation in which they dress up as a baby or young child and place themselves under the control of an 'adult'. Accordingly, such sex games frequently involve imaginary wrong doings to create excuses for chastisement and reinforce the dominance and power of the parent figure.

Clare Mansfield – or 'Mistress Chloe' as she is better known to her clients – provides an ideal example of pseudo-mother. Dominatrix and purveyor of painful bliss at Fettered Pleasures on the Holloway Road in London, 45-year-old Mansfield is a professional supplier of masochistic services, or as it is now called, 'BDSM' – bondage, domination, submission, masochism. As Mansfield freely admits, the infantile origin of BDSM is undeniable – to her clients, who include politicians, film stars and captains of industry, Mistress Chloe is purely and simply a pseudo-mother. When asked, for instance, about her attitude towards those who she canes, whips or simply mentally humiliates in her 'dungeon', Mansfield insists, 'You can't hate men in this business. You have to love them. I'm a mother figure.' Like perpetual children who desperately crave a dominant figure to satisfy their infantile insecurity, visitors to Fettered Pleasures derive intense comfort by playing out scenarios that similarly involve a super-dominant figure being there to protect them.

Masochists are not 'different', they are merely – like homosexuals – further down a continuum of infantile behaviours that are carried through into adulthood. Since an infantile dependence on our sexual partner very much defines human relationships, it could even be argued that they are more highly evolved. In the same way that homosexuality results from a greater extension of one infantile behaviour – a childlike desire to form relationships only with members of the same sex – so masochists retain a more intense version of an alternative childlike behaviour, which is a desire to seek domination and thereby protection and comfort from another individual, male or female. For the so-called 'normal' rest of us, we seek comfort in a slightly milder form by seeking cuddles, kisses and doing what we are told by our sexual partner.

OEDIPUS AT HOME

While the mother–infant relationship pervades every aspect of our sex lives, it has perhaps been most acutely identified in the form of what Freud referred to as the 'Oedipus conflict' – that is, the stage that human males go through when they develop an unconscious sexual desire for their mother, while also envying and fearing their father. Put another way, it is the emergence of a competitive feeling in the boy as he strives to oust his dominant father from the position of head of the household. Oedipus, according to Greek mythology, was the son of Laius and Jocasta who – because he was raised by a shepherd and was unaware of his true identity – happened to kill his father in a fit of road rage, and then unwittingly married his mother. When the truth was revealed, Oedipus proceeded to blind himself, while Jocasta hanged herself.

According to Freud, the Oedipus conflict is a state of anxiety that should naturally disappear during childhood as the boy gradually takes on the image of his father, except in some individuals in which it persists and causes problems for their relationship later in life. Needless to say, in a species that has had its period of infantile dependency extended as far as ours has – that is, considerably beyond sexual maturity – it is little wonder that feelings of maternal dependency and sexual desire occasionally get more than a little confused. Remember too the research of psychologists at St Andrews University, which suggested that men choose partners who resemble their mother, while women choose partners who resemble their father. What is more, because of the father's dependency on his wife to act as a pseudo-mother, it is for good reason that he and his son may develop strong competitive feelings towards each other. Since both are competing for the maternal affections of the boy's mother, it is perhaps more akin to sibling rivalry than straight father–offspring conflict. However much a father loves his child, it is inevitable that he will feel jealous over its power to demand love and attention from his own pseudo-mother. To add insult to injury, the young child even has liberal access to the part of his sexual partner's that was previously reserved for him alone – her breasts.

Because a father–son conflict, as well as a strong maternal dependency by sons, is well recognised in various cultures, there exist many elaborate and often traumatic rituals to drive sons away from their mothers and force them to recreate mother–son relationships with other women. An example of this is provided by the Samba tribe of Papua New Guinea, in which fathers and other men belonging to their generation subject boys as young as seven years old to a series of 'punishments'. During these rituals, the boys are repeatedly beaten around the head by the adults causing blood to stream from their noses, and if the boys cry or attempt to resist then the adults simply prolong and increase the punishment. Observers of these rituals describe how the terrified boys cower, fearing for their lives, during this brutal assault. But to all involved, the symbolism of the nose bleeding is clear since the Samba equate the nose with the penis – it is an unambiguous warning that the adults will react unmercifully towards any boys who show the slightest sexual interest in women belonging to the older generation. By way of reinforcement, such rituals continue until, at the age of sixteen, the boys are told that they may never so much as touch, hold, talk with, eat with or look at their mothers. Should they disobey, they are warned, they will be killed. As University of Chicago anthropologist Gilbert Herdt pointed out, the boys 'must be traumatically detached from their mothers and kept away from them at all costs'.

Like other strong human urges, including drug-taking and infidelity, incest often has to be countered by the most extreme prohibitions. According to Freud, the necessity for all human societies to create strict taboos against incest is because such an inhibition 'is not to be found in the psychology of the individual'. Among the most draconian of deterrents against incest are those of the Cayapa Indians of Ecuador who suspend anybody who is found guilty of incest over a table that is covered with candles and roast them to death, while the Incas gouged an offending man's eyes out and then cut his body into quarters. On the other hand, in some societies there is such shame associated with incest that anybody who commits this crime feels obliged to inflict a severe punishment on themselves – for example, the native American Kaska tribe express their remorse for committing incest by leaping into fires or tearing off their penises. One investigation into incest around the world found that 31 per cent of the societies studied punished incest with death, mutilation, sterility or expulsion from the community. The occurrence of such taboos clearly contradicts the view that members of our species have a universal aversion towards having sex with relatives – or at least those who we are brought up with, as proposed by Freud's rival Edward Westermarck – while supporting the view that our species has a tendency to be attracted to close family members.

Incest, be it mother–son, father–daughter or between siblings, is a characteristic of our species that permeates every society and economic level. Dr Ruth Weeks, a child and adolescent psychiatrist, noted that people who commit incest include 'judges, ministers, university professors, doctors, teachers, skilled workers, white-collar workers, farmers, and unskilled laborers'. In a review of research into incest, Blair and Rita Justice found one study that stated that 4 to 5 per cent of the American population is involved in incest, and another that concluded that 'incest implicates at least 5 per cent of the population and perhaps up to 15 per cent'.

However, of all the types of incest, father–daughter incest is by far the most common. While acknowledging that a significant proportion of such cases are simply fathers looking for an outlet for their sexual urges – with the daughter merely being chosen because of her proximity – psychological profiles of offenders suggests that there is often more to it than this.

Men who have been convicted of incest typically have three qualities in common – each of which points towards a failure to develop beyond the infantile Oedipal stage. Firstly, they never identify with their father. Secondly, they are treated as 'little men' by their mothers and take responsibility for looking after the household. And finally, they have a persistent fixation with their own mother. All in all, such boys perceive themselves as rightful inheritors of their father's position, and therefore assume that they should naturally receive the lion's share of their mother's sexual as well as emotional favours. For this type of male in particular, the transfer of their emotional dependence from their natural mother to a pseudo-mother, in the form of a wife, can be extremely difficult. Even if such men do manage to make a partial break and get married, their wives are highly likely to suffer from constantly being compared with their husband's mother. In such cases, one possibility is that they turn to their daughter – who like their wife, but unlike their mother, is their own flesh and blood – to act as a replacement for their lost mother. Such a theme runs through the background of a great many incestuous fathers, especially in situations where sexual relations have broken down between husband and wife.

In situations where the daughter is less than fully adult, father–daughter incest is morally reprehensible and doubtless a form of child abuse. Children are not psychologically well enough developed to appreciate the consequences of their actions, while the father, for whatever motivational reasons, is abusing his daughter and the trust that she places in him. Nevertheless, while different authors fervently disagree on the proportions of willing and unwilling daughters, it does seem that small proportion of daughters – especially those who are older – actively seek to initiate a sexual relationship with their father. Blair and Rita Justice recount an example of such a situation in their book *The Broken Taboo*:

> Barbara was hungry for affection, for approval. She tried to win her mother's love but she felt she never could. She would try to help do the ironing but her mother would tell her to go away. The mother never seemed to have time for her or interest in her. With her father, she got the affection and approval in the form of sexual activity. She took over from the mother in terms of supervising the other children. She became the lady of the house, right down to being her father's sexual partner.

To understand the basis of this extreme form of 'father–love' – or what psychologists Evelyn Pitcher and Lynn Schultz refer to as the 'early coquettish overtures' of daughters – it is important to remind ourselves of the opposing goals of human male and female love. Whereas male passion responds strongly to a nurturing mother–figure, female passion responds strongly to childlike dependency in adult males. As mentioned previously, one illustration of the intensity of this drive is the occurence of young girls becoming besotted with large powerful horses. In situations where husbands have apparently been abandoned by their wives and are in need of support, it is possible to see how daughters could respond by providing them with the type of care and support that would normally be provided by his wife. There undoubtedly exists a strong link between a socially and emotionally disturbed relationship between husband and wife, and incest. In time, and especially if the daughter derives increasing satisfaction from her ability to provide for her father's needs, it is quite possible that a desire to 'complete' her role as a mother–substitute could cause the daughter to accept or even actively maintain the sexual relationship. In one study, psychologist Herbert Maisch found that 23 per cent of the female victims of father–daughter incestuous relationships actively encouraged the situation to continue. While the 'little-mother' explanation for father–daughter incest should not be overstated, it is fair to say it does play a significant part in a limited number of cases.

All in all, and regardless of each individual society's mores, it appears that our species' evolution has significantly contributed towards the likelihood of incest occurring. With males seeking partners who can satisfy their lifelong need for mothering, and females directing their maternal care towards adult males, incest has become more of a possibility than perhaps ever before. While in no way vindicating such activities, this explanation merely underlines the fact that our species is in a uniquely precarious position with regard to the development of 'normal' – as exists in other non-infantised species – sexual behaviour.

PAEDOPHILIA

If incest is widespread, then paedophilia exists in epidemic proportions. Despite being abhorred by Western society in particular, paedophilia is staggeringly common throughout the world. A recent survey of children's sexual encounters with adults provided the following account of cross-cultural attitudes and behaviour:

Ethnographic reports show that many non-European cultures have few inhibitions about stimulating rather than suppressing children's sexuality. Mohammed himself had a child bride, and in Muslim societies girls can become betrothed or actually married and having intercourse at ages we should consider incredibly premature. Up to 1929, girls could be legally married at 13 in England. According to Edwardes & Masters (1970) Muslim boys often had their penis tweaked or jiggled by their mother, nurse or other attending female, such being the custom. They also cite (p. 125) descriptions of Swahili fathers in Zanzibar massaging the penis of their little boys of 4 to 6 to stimulate growth and ensure potency. Sex initiation rituals for pubertal boys were sometimes of a homosexual nature. In the Kaluli tribe of New Guinea, for example, boys of 10 or 11 would be introduced by their fathers to a man who would become for some months their training partner for penetrative anal intercourse.

Meanwhile, a report on the history of male homosexuality in Muslim societies states: 'That all men were susceptible to boyish beauty was taken for granted. Many would desire sex with them and some would do it. The Habalite jurisconsult Ibn al-Gauzi wrote: "He who claims that he experiences no desire [when looking at beautiful boys or youth] is a liar, and if we could believe him, he would be an animal, not a human being".'

Today, the problem of paedophilia is simply enormous, and the vast majority involves men who prey on young boys for sex. Two-thirds of all sex offenders in US prisons are convicted for child sex abuse, and according to a variety of surveys, it is estimated that somewhere between around 8 per cent of all boys and 21 per cent of all girls in both the USA and the UK are sexually abused. In a review of various surveys into child sex abuse, David Finkelor – Director of Crimes against Children Research Center, and Professor of Sociology at the University of New Hampshire – found that estimates of prevalence vary from 3 to 30 per cent for males, and from 6 to 62 per cent for females. Moreover, 90 to 95 per cent of all paedophiles are men.

So, what is going on? What causes so many men to be sexually attracted to children? One clue comes from the fact that paedophiles are far more likely to target those of the same sex than would be expected. Whereas only 3 per cent of the general male population choose to have same-sex relationships, among paedophiles the figure rises to around 33 per cent. There is also a strong tendency among paedophiles to socialise only with children, reject relationships with adults, and choose professions that are centred around children's lives and activities, including childcare or teaching. In essence, paedophiles frequently retain an inordinate number of infantile behaviour patterns into adulthood – including that of wanting to associate only with members of the same sex, hence the disproportionate number of paedophiles who exclusively target young boys. Contrary to popular opinion, the typical paedophile is anything but a 'rabid sex fiend', and instead is more likely to be religious, moralistic, guilt-laden and lonely. It is highly significant that, of all paedophiles, those who are the most committed to children as sex objects and who find it most difficult to transfer their sexual attention to adults, are also those who are at the most infantile end of the scale – in other words, they are poeple who not only want to limit their contact to children, but also remain at the early stage of preferring to associate only with members of the same sex. Paedophile men who target young boys are not only far more likely to restrict their sexual activity only to children, but also twice as likely to reoffend as those who target young girls.

A number of psychologists have similarly proposed that a failure to develop fully is at the heart of paedophilia. Anthony Storr, for example, commented:

When sexual impulses are denied their normal fulfilment in an adult love relationship they continue to seek expression in ways which are generally abandoned by those who have been able to reach a more mature stage of development. Very many people have been paedophilics in one sense; for very many people have had sexual contacts with children when they themselves were children. When adult sexual expression has never been attained, the desire for sexual contact with children may persist.

Similarly, C.K. Li and others wrote:

To the extent that adult sexual perversions resemble childhood pregenital activities, they are often understood in terms of either a regression to, or a fixation at, a pregenital (hence immature) stage of development.

While it sometimes happens that people who undergo psychoanalysis are initially surprised by the explanations given for their particular problem, in the case of paedophiles the subjects are often all too aware of the infantile origin of their particular sexual obsession. A case in point is Andy, a 34-year-old unemployed man, who gave the following explanation for his sexual behaviour during an interview with a psychologist:

Children will pick up, they will learn sex with other boys, do it to each other, and this is what I found out, they do it to each other. And when you get older, you still fancy smaller kids, because in your age, you haven't really made up your own mind, your previous experiences when you first started. So, what I'm saying is you may be fifty-odd, right, but you still fancy small children because the men haven't actually grown up, grown out of the sexual ways when they were kids themselves, I think this might be the problem. The kids, still in their kiddy minds, fancy other kids, and I think this is really what it is – it is with me … What I was saying here – mmm – I think, as I grew up, I don't, I don't think sex come along with me at the same time, I learned that at a much later date … I learned sex when I was about 19.

For other paedophiles, their infantile nature is simply expressed as a difficulty with adult relationships. In one UK survey of paedophiles, more than half of the respondents expressed the view that a difficulty in relating to adults was associated with their sexual preferences. Occasionally, paedophiles actually describe themselves as having a strong hatred for adults and wanting to seek the company of children at all times.

In the light of paedophiles' child-like qualities, it is should come as no surprise to find a strong link between paedophilia and the Church. Just as with the link between homosexuality and religion, paedophilia's infantile origins mean that those who display this behaviour will also possess a particularly strong need to be 'mothered' or 'fathered'. In the same way that children crave the comfort and protection offered by their parents, so those at the infantile end of the scale, including paedophiles, are most likely to crave the comfort and reassurance that religion offers. To say that child sex abuse is rife in the Catholic Church would be no overstatement. Moreover, a predilection of Catholic priests for sex with young children is nothing new. In 1931, a secret report in Ireland denounced the 'alarming amount of sexual crime' by priests, involving many 'criminal cases with girls and children from 16 years downwards, including many cases of children under 10 years'. Recently, Cardinal Bernard Law, who leads the Boston diocese of the Catholic Church, handed over the names of eighty priests who had been accused of abusing children, while the *Boston Globe* newspaper has estimated that the compensation claims brought by hundreds of alleged victims over the last decade could exceed $100 million.

In Ireland, fifty-two 'industrial schools', for children whose parents were too poor to look after them, were closed in the 1970s amid allegations that priests and nuns subjected most of the children in their care to physical or sexual abuse. Meanwhile, in England numerous priests are serving prison sentences for child abuse, including Father Adrian McLeish who abused four boys – the sons of parishioners – and who was found to have the UK's biggest collection of child pornography in his home. In March 2002, even the Pope was forced to speak out against what he called the 'sins of our brothers' while admitting that some priests had succumbed to 'the most grievous form of evil'.

To conclude, far from being 'selected for' in evolution because of its advantages for the individual, paedophilia is yet another accidental by-product of our species' regression into a permanently infantile state. In the same way that a child will actively seek out other children to be with, and react timidly in the presence of adults, so will a paedophile. As with incest, this evolutionary explanation for paedophilia is not to condone it in any way, but merely to see it as a devastatingly sad and unfortunate consequence of our species' infantile nature.

TRANSVESTITES AND TRANSSEXUALS

Being a *berdache* – that is, a man who behaved much like a woman, or occasionally a woman who behaved like a man, in a Native American tribe – was no soft option. As well as customarily providing non-berdaches with sex, berdache men carried out a variety of shamanistic duties, became skilled at women's tasks, wore women's clothes, and even underwent their own versions of menstruation and childbirth. After a male berdache got married – which was usually to a young man or an older one who was between marriages – he began to imitate menstruation by scratching himself between his legs with sharp sticks until blood flowed, after which he strictly followed the women's menstrual taboos. When he decided to become pregnant, he stuffed rags and bark under his skirt in increasing quantities, boasted of his condition in public, and started to follow normal pregnancy taboos – except that, unlike pregnant women in the tribe, he allowed his husband to carry on having sex with him. As delivery approached, the berdache started to drink a concoction of mesquite beans that led to extreme constipation, pain and stomach cramps – in other words, 'labour pains'. Finally, and inevitably, when the colonic pressure became too great, the berdache went into the bushes, took the position of a woman giving birth, and 'gave birth'. Clearly, the 'baby' simulation had insummountable problems from this point on. As a result, the faecal baby was pronounced stillborn, following which the berdache commenced a period of wailing to lament the loss, clipped his hair and gave away the cradle that he had prepared for his child. The husband, meanwhile, was obliged to join in with the melodramatic proceedings.

In the same way that homosexuality, paedophilia and incest are universal features of our species, so too are transvestism and trans-sexualism – conditions in which individuals feel the urge to adopt the clothing, or possess the body, of the opposite sex. Throughout the world, there exist many examples of men who behave like women, from the *kathoey* of Thailand, to the *xaniths* of Oman, the *hijras* of India, the *mahus* of Polynesia and the *hsiang ku* of China. There are also many cultures in which it is perfectly acceptable for certain women to behave more like men, such as the 'Amazon' warriors of the Tupinamba Indians of north-eastern Brazil. When the explorer Pedro de Magalhães de Gandavo visited that region in 1576, he gave the following account of what he observed:

> There are some Indian women who determine to remain chaste: these have no commerce with men in any manner, nor would they consent to it even if refusal meant death. They give up all the duties of women and imitate men, and follow men's pursuits as if they were not women. They wear the hair cut in the same way as the men, and go to war with bows and arrows and pursue game, always in the company with men; each has a woman to serve her, to whom she says she is married, and they treat each other as man and wife.

In a species that has had its development curtailed to the huge extent that ours has, gender confusion is hardly surprising. With human development taking such a long time to be completed, we each go through an astonishingly long and vulnerable period during which any number of factors can upset the normal sequence of events. Bear in mind, for example, the link between low levels of the sex hormone testosterone and marital stability, and thus the clear benefits of females selecting males with low levels of this hormone which is essential for the development of gender characteristics. In addition, the possibility has already been mentioned that the brains of those at the infantile end of the scale are essentially less sensitive to developmental hormones than the rest of the population. All these factors, along with the prevalence of gender confusion in our species, suggest that a major consequence of our species' infantile evolution, has been to leave many of us precariously balanced on the edge of failing to develop completely as one sex or the other.

For those at the infantile end of the human spectrum, development towards becoming a boy or a girl will be more gradual and tentative. As discussed earlier, it is a situation that can lead to homosexuality, through some individuals failing to develop beyond the juvenile phase of seeking relationships only with members of the same sex. There are, however, a variety of other possible effects of a highly extended infantile period that may or may not combine with homosexuality – the precise combination of these effects depending on both the genetic make-up of the individual and the environment. In situations, for example, where a boy or a girl takes longer to develop the desire to join single-sex groups, or where their environment makes it impossible to join such a group, there is the distinct possibility that they will adopt one or other of their parents as a role model, and essentially imprint on them regardless of sex. Without the reinforcement of same-sex peers, the situation could therefore arise where children can mimic and retain many behaviours of the opposite sex. With mothers being the dominant presence in most children's lives, clearly this places boys at the greatest risk of such 'cross-identification', and accordingly, transvestism is very rare in women. As Arlo Karlen points out:

A boy must wrench himself away from the first object of his love and identification in order to become a man. A woman's final identification is the same as her first – feminine. This is probably why masculinity is more vulnerable to threat, breakdown and distortion than femininity; and therefore why more men than women in every society studied for such things show extreme casualties in psychological development.

A further reason for the greater vulnerability of the male is the greater infantising of the males than females, but more of this later. Homosexuality – in which there is no cross-identification – is evidently just one version of events that can arise from a particularly infantile development. If a limited amount of cross-identification combines with homosexuality, then the result is effeminate behaviour in gay men – giving rise to 'drag queens', 'she-males' and 'female impersonators' – or masculine behaviour in lesbians. In the most extreme situation, however, in which cross-identification has occurred to the greatest degree, the result is somebody who identifies so strongly with the opposite sex that they feel compelled to change the sex of their body, in which case they become 'transsexual'.

In contrast, there are also people who experience cross-identification and yet develop beyond the same-sex group stage to become heterosexual. In the same way that homosexuals need not experience any cross-identification, so transvestites need not be homosexual. Arlo Karlen notes, 'There is no simple formula for the various degrees and combinations of cross-identification, effeminacy in men and masculinity in women, homosexuality, and transsexualism. Most of these phenomena can occur alone or in various combinations.' However, from drag queens in full make-up, wigs and flowing skirts, to happily married men who covertly wear ladies' underwear, and from June Buckridge to Childie in Frank Marcus's *The Killing of Sister George*, what unifies this extraordinary collection of behaviours is that they ultimately share an infantile origin.

But despite the mind-boggling variety of appearances and sexual expressions that our species displays, since the greatest variety occurs among the most infantile members, there are certain recurring features within this group in terms of their non-mature qualities. Transvestites – like homosexuals – display a wide variety of essentially youthful and dependent characteristic, and those who combine both homosexuality with transvestism are likely to have the most immature personalities of all.

The link between transvestism and infantile dependency is particularly apparent within the church. The gender-ambiguous nature of religious garments was perhaps most memorably, if irreverently, expressed by the American actress Tallulah Bankhead, who is said to have remarked to a robed thurifer – a person who carries incense during religious ceremonies – 'Love the drag, darling, but your purse is on fire.' More recently, the connection was noted by Boy George, the pop icon of transvestites, when he wrote 'I dress in a similar way to a priest or an archbishop ... I wear robes, not dresses, and to be a transvestite you must wear women's undergarments. I don't.' Marjorie Garber, Director of the Center for Literary and Cultural Studies at Harvard University, has extensively documented the association between transvestism and the Church. She points out, for example, that the female transvestite saints of the Middle Ages were numerous, despite the Deuteronomic prohibition against wearing the clothes of the opposite sex – the most famous being St Joan – and that the male clergy were among the first in England to adopt the wig and parasol – accessories that were previously used only by women. Gerber notes that 'particular [religious] items of clothing have tended to cross over gender lines, not through uniformity *per se* ... but rather by the migration of styles over time from one gender to another'. Throughout the ages, priesthood has been equated, often with great derision, with effeminacy. Describing the Conclave of Cardinals in the Vatican, former nun Patricia Hussey commented, "Rows and rows and rows of old white men in lace dresses were watching the cardinals make their entry.' As with homosexuality, it appears that transvestism too has a long and honourable association with the Church.

Alongside childlike dependency comes an equally childlike talent for mimicry. Accordingly, there is a long history of transvestites excelling in the arts. Of course, to a certain extent play-acting is at the heart of a transvestite's expression of his or her cross-gender personality. Some of the most famous transvestites in history have been 'actresses' and 'comediennes'. Although the all-male kabuki tradition in Japan is perhaps the best-known example of an Asian transvestite theatre, similar acting traditions have existed throughout the world. The transvestite acting tradition forms the basis, for example, of Balzac's story in which Sarrasine, a young sculptor, becomes infatuated with a Roman opera singer called La Zambinella, only to find out later that 'she' is actually a castrato. In all fairness to Sarrasine, the accuracy with which transvestites impersonate women, from their physical appearance to minute body movements, can be astonishing. In contemporary popular culture, cross-dressing and 'androgyny' have become almost de rigueur, as evidenced by David Bowie, Boy George, Twisted Sister, and the New York Dolls. Dee Snider, male lead singer of Twisted Sister, was even given the accolade of being voted one of the worst-dressed women of the year in 1984.

The mimetic talent of those inclined to cross-dress has also led to an association between transvestism, diplomacy and espionage, with one of the most famous transvestites in Western history being the Chevalier d'Eon, a diplomat and later an actress, of whom it was written, 'At d'Eon's death [in 1810], according to some calculations, he had lived 49 years as a man and 34 years as a woman. The English court, and the French king, "proved" he was a woman. And the autopsy "proved" he was a man.'

The final link between infantising, transvestism and transsexualism comes from the most infantile group of humans of all – the Mongoloids. As mentioned at the beginning of this section, berdaches are characteristic of most if not all the Native North American tribes – these being people of Mongoloid origin who reached America from East Asia about 15,000 years ago. But even more impressive, at least considering the contemporary situation, is the conspicuousness of male transvestites and transsexuals – as well as homosexuals – in countries such as Thailand. Moreover, the extreme infantising of the Mongoloid people has also resulted in the males and females having remarkably similar body forms. Since Mongoloid males lack many obvious signs of maturation – such as extensive body hair, coarse skin, a heavy jaw and thickened skeletal bones – there exist far fewer physical differences between them and females than in other races. It is a feature of Far Eastern males that has caught out many an unguarded foreign visitor, as the following passage from the American press in 1986 vividly illustrates:

A former French diplomat and a Chinese opera singer have been sentenced to six years in jail for spying for China after a two-day trial that traced a story of clandestine love and mistaken sexual identity ... M. Boursicot was accused of passing information to China after he fell in love with Mr. Shi, whom he believed for twenty years to be a woman.

Beyond a shadow of a doubt, we are an extraordinary species. After millions of years of being infantised, our sexual behaviour is now a complex mixture of mature and infantile elements. While sex is ubiquitously linked to infantile insecurity, for some the relics of childhood are even more pervasive. From masochism to incest, and transvestism to paedophilia, a wide variety of our most curious present-day behavious can ultimately be traced back to our species' irresistible retreat from maturity. They are, once again in evolutionary terms, prices to be paid for our unparalleled social and intellectual talents.

chapter nine

vive la différence

That our species is infantile is beyond doubt. But what is equally certain is the fact that you and I differ in the amount that our infantile behaviours have been retained into adulthood. So what causes some of us to 'grow up' more than others? This chapter will be looking at the four major factors that influence how mature, or infantile, each of us ends up being. These factors are our sex, our racial origin, the social strategy of our ancestors and our upbringing.

SILLY BOYS AND SENSIBLE GIRLS

In 1967 there appeared an organisation known as SCUM, the Society for Cutting Up Men, and its manifesto contained the following comments and resolutions:

To overthrow the government, eliminate the money system, institute complete automation and destroy the male sex ... To be male is to be deficient, emotionally limited; maleness is a deficiency disease and males are emotional cripples. The male ... is a half-dead, unresponsive lump, incapable of giving or receiving pleasure or happiness ... SCUM will kill all men who are not in the men's auxiliary of SCUM. Men in the men's auxiliary are those men who are working diligently to eliminate themselves ... men who kill men.

Although largely forgotten today, SCUM did manage to achieve some notoriety when on 3 June 1968, its founder, actress Valerie Solanas shot and seriously wounded the artist Andy Warhol. While it was unquestionably an ultra-extreme feminist organisation, what is relevant about SCUM is that it did find considerable support among lesbians because of its anti-male stance. Although hardly representative of every lesbian's sentiments towards men, the support for SCUM does illustrate how an enmity and distrust of the opposite sex is considerably more prevalent among female than male homosexuals. It is far less likely, for instance, that a gay pressure group should want to rid society of women. But why? What are the fundamental differences between the sexes that, in addition, make men three times as likely to become homosexual, as well as significantly more likely to suffer from mental as well as growth disorders than women?

Although members of both sexes vary in the degree to which they retain infantile characters into adulthood, males lag behind on just about every score of development that exists. Evidence that human males are more infantised than human females comes from their bodies as well as behaviour. It is well established that the Y-chromosome, the genetic material that specifies that an individual will develop as a male rather than as a female, has the effect of slowing down the rate of development. One consequence of this is that male babies are typically born more immature than females babies, and as a result 5 per cent more male babies die at and around birth than female babies. It has also been discovered that male babies develop slower in the womb than female babies. Unborn baby boys are, for instance, four to six weeks behind girls in their bone development. Boys also move their mouths later and less frequently than girls before birth, this action being an essential skill that allows a baby to breastfeed as well as make rudimentary sounds. Immediately after birth, newborn girls continue being ahead of boys in many aspects of their development – in fact, the same trend continues throughout childhood, with boys consistently having more extended infantile stages than girls. One consequence of boys having a more protracted infantile development is that men's brains grow to a slightly larger size than those of women. It is not a large size difference – a mere 100 grams, or roughly 7 per cent of the total brain size – and it is certainly meaningless in terms of intellectual ability for reasons given earlier, but it nonetheless exists even when the size difference of men and women is taken into account.

When it comes to the emergence of the single-sex gang, boys once again lag behind girls by at least one year. While three-year-old girls show an increasingly strong preference for playing among themselves, three-year-old boys are far more solitary in their play and if anything reject interactions with other boys. At this age, boys still gravitate towards a mother-dominated world and prefer the domestic role-playing of girls in which they are invited to be passive recipients of female care. Although it does not last for long, one consequence of this late-development of boys is that around the age of three, girls are dominant to boys. Girls at this age are typically more confident than boys, as indicated by the fact that boys suffer considerably more 'separation anxiety' when they first go to school.

As children head towards adolescence, boys remain significantly behind girls in their growth and social development. Around the age of eleven or twelve, many girls suddenly become taller than boys, the reason being that boys' growth spurt takes place roughly two years later than that of girls. Puberty, too, begins about two years later in boys. The more immature and playful nature of the human male goes a long way in explaining why boys constantly fall behind girls in their academic achievement. In England, for example, boys are perenially around nine per cent below girls in their ability to achieve a 'c' grade or better in their mid-teenage examinations. In virtually every aspect of their lives – from growth to puberty, educational aptitude and social development – human males lag considerably behind females, having had each of their infantile stages extended that much further into their lives.

One result of having such a slow development is that boys are far more at risk from disturbances, both physical and emotional, during childhood. In his book *Man and Woman*, the medic and essayist Henry Havelock Ellis provides a catalogue of medical disorders that males suffer from more than females. The following are excerpts from a long and rambling text:

Again, in London, at St. Thomas's Hospital, if we take, for instance, the years 1881–87, we find that hare-lip, for example, was found in 43 males to 20 females. Bryant's note-book (according to Braxton Hicks) showed 44 males to 20 females, almost the same proportion, while Manley found 27 males to only 6 females. Double hare-lip is almost exclusively found in males ... Nearly every other important form of malformation is found more frequently in males than in females ... Supernumerary digits are much more frequently found in males than females. Supernumerary breasts and nipples are also commoner in men; among several hundred persons Mitchell Bruce found that 9.1 per cent. men and only 4.8 per cent. women possessed an extra nipple, so that it is nearly twice as frequent in males ... The majority of muscular abnormalities are found in male subjects ... [And so on.]

Havelock Ellis's subsequent observation that mental disorders are even more marked in their prevalence in males has been confirmed by recent research. It has been found that men are, for instance, five times more likely than girls to suffer from delinquency, three to five times more likely to suffer from attention deficit disorder, and three times more likely to suffer from learning disabilities. They are also much more likely to exhibit behavioural problems after emotional disruption, such as divorce of their parents. Together with the information on the slow rate of development in the human male, all these findings point towards human males being far more infantised and fragile in their development than human females.

Men being more infantile than women explains a great deal. As well as explaining why men often have a strong proclivity for 'messing around' and spending time on less than useful pursuits – in terms of their family commitments at least – it also partly explains why homosexuality is roughly three times more common in men than women. Since men take longer to develop, and have more extended childhood periods, they have a greater chance of failing to develop

beyond the same-sex phase. But there is another fundamental difference between the two sexes that affects the social behaviour of men
and women, and further contributes towards the differences between
lesbians and gays. Moreover, it is a difference that goes back a long
way in our evolutionary history.

As mentioned previously, humans have a breeding system that
typically involves a practice called 'female exogamy' – that is, males
stay in the territory into which they are born, while females move
away to find a sexual partner in a nearby territory. Exactly the same
pattern exists in our close relatives, the chimpanzees and bonobos, in
which males stay in the troupe into which they were born, while
females transfer to another troupe soon after they reach puberty
around the age of eleven. The effect of this practice on the social
behaviour of the two sexes is massive and far-reaching. For males, it
makes it imperative that they establish their place in the social hierarchy as soon as possible, since this will affect their mating success later
in life. They need to form close alliances with other males to raise
their place in the hierarchy. For females, on the other hand, childhood and early adolescence is primarily a time when it is best to keep
their heads down, keep out of harm's way, and simply survive until it
is time for them to leave and form long-lasting adult relationships in
another group. The advantages of females forming relationships with
numerous other members of the group are minimal since they will
soon be leaving the group. Whereas young males typically form
'extensive' political relationships with a large number of other males,
young females typically stay far closer to their mothers and form a
small number of 'intensive' relationships with other young females.
Perhaps most importantly, it is vital for a young female to keep well
away from adult males in her home group – not only because of their

frequent aggressive and dangerous behaviour, but also because of the fact that they are all closely related to her and she needs to avoid the possibility of an incestuous mating. For adolescent female chimpanzees and humans alike, it is essential that they are extremely cautious of close relationships with adult males at a time in their life when they become sexually receptive but have yet to move away from the group into which they were born – far better to stick to close relationships with other females for the time being.

While it is clear how this feature of our species can have a strong influence on the behaviour of boys and girls, how does it relate to the behaviour of adult homosexuals? The answer can be found by being more precise about the infantile origin of homosexuality. Rather than simply retaining an excess of 'immature' behavioural patterns into adulthood – homosexuals can more accurately be said to retain an excess of 'immature behaviours that are appropriate to their sex'. Homosexual men, in other words, retain an excess of immature male behaviours, while homosexual women retain an excess of immature female behaviours. From this viewpoint, it can now be seen that – since immature males form stronger same-sex relationships than immature females – adult males who retain immature behaviours into adulthood will form stronger same-sex relationships than similarly infantised females. Alongside the human males' more infantile nature, this provides a further explanation for why there are more male than female homosexuals in every known human society.

Furthermore, it also provides an explanation for 'SCUM effect' – that is, lesbians having a reputation for being more cautious of members of the opposite sex than is the case with gays. One reason for this is that, at the stage when girls form single-sex relationships –

in other words, before they have normally left their birth group – it is vital for them to avoid close contact with mature men because of the high chances that these males will be closely related to them. Whereas maturity in females brings with it an inclination to form relationships with males, immaturity is associated with keeping well away from them. In contrast, young males have no such concern.

This, however, is not quite the complete picture with regard to differences between homosexuality in human males and females. Homosexual men and women also differ hugely in the average number of sexual partners that they have. Alan Bell and Martin Weinberg, from the Kinsey Institute for Sex Research, found that the average gay man has thirty-three times the number of sexual partners than the average lesbian, while other studies have claimed even larger differences than this. The Kinsey researchers also found that nearly half of the gay men had over five hundred different sexual partners, another third had between one hundred and five hundred, and over 90 per cent had at least twenty-five sexual partners. Much of this sex had taken place between comparative or absolute strangers. Lesbians, on the other hand, were found to be much more like heterosexual women than homosexual men in this respect. Most were found to have had fewer than ten female sexual partners, and comparatively little sex had taken place between strangers.

So why do many male homosexuals have such a vast sexual appetite? The answer comes in two parts – one is to do with being infantised, the other is simply to do with being male. To begin with, in being more infantised, homosexuals possess a greater amount of playful and experimental behaviour than other males. In a childlike fashion, they are more driven to seek variety and are quick to move from one experience to another. Although heterosexual males pass through a similar phase early in their sexual career, they are considerably more likely to develop more mature reproduction-oriented behaviour as they grow older. Homosexual males, on the other hand, remain in the more youthful experimental phase of sexual development for far longer. This is not to say, of course, that many gay men do not form long-lasting relationships sooner or later. In various surveys, between 40 and 60 per cent of homosexual men questioned were currently involved in a steady relationship. Indeed, the pop star Boy George is quoted as saying, 'There's this illusion that homosexuals have sex and heterosexuals fall in love. That's completely untrue. Everybody wants to be loved.' However, to this it could be added that homosexual men typically take longer to settle down than others.

Edward Wilson, Harvard professor and the father of sociobiology, provides the second reason for the marked difference in gay and lesbian sexual behaviour. 'It pays for males to be aggressive, hasty, fickle, and undiscriminating [while] it is more profitable for females to be coy, to hold back until they can identify males with the best genes.' For men, the act of copulation is almost a study in efficiency. With minimal foreplay required, a few thrusts and 100 million sperm per ejaculate that are merely stripped-down sacks of DNA with a tail attached, mating is anything but costly in terms of energy invested. Most importantly, once a male has finished copulating, that is the end of it, and there are no further physical consequences. For females, on the other hand, this could not be further from the truth. While the act of copulation itself is admittedly low in its energy demands, the potential after-effects can be quite astronomic in their energy drain for years to come – that is, if fertilisation occurs and the female becomes pregnant. Females are therefore much more cautious about mating than males. What is more, if the female of any species should be excessively careful, it is the human female, since she devotes more time to raising her young than any other species. It is an incredibly dangerous risk for a women to have sex at random. Without some sort of quality control she could end up spending years raising an extremely unhealthy child who is likely to die at any time.

So, with men and women having such contrasting approaches towards sex, heterosexual relationships must be something of a *modus vivendi* – a compromise. Even taking into account the infantising of our species and the evolution of pair-bonding love, strong differences nevertheless exist between women and men's sexual behaviour. Men will inevitably tend towards having sex with more partners and prefer less courtship and commitment, while women will encourage a longer courtship period and more commitment. Contraceptives aside, such a conflict has existed since time immemorial. In contrast, no such conflict exists in the majority of homosexual relationships. Homosexual men, rather than having to contend with reluctant female sexual partners, now have male sexual partners that are just as enthusiastic about opportunistic sex as they are. Similarly, homosexual women now have female sexual partners that are likely to be as careful, and choosy, about their sexual partner as they are. As Simon Le Vay points out:

Lesbians present you with an opportunity for studying female sexuality in its purest, most undiluted form, and gay men for studying pure male sexuality. With heterosexuals, if you're trying to distinguish how men feel and how women feel sexually, the waters are muddied because perforce you're mixing the two … With gay people, however, the drive is the same in both partners, so they have a much better chance of finding partners interested in the same frequency of sex and the same variety of partners as themselves.

In conclusion, beside the fundamentally different reproductive strategies of the two sexes, the greater infantising of the human male accounts for many of the differences in the biology of human males and females. In all examples of gender variation – be it homosexuality, transvestism, transsexualism, or any other departure from the norm – males predominate. Males possess a greater susceptibility to mental, physical and physiological abnormalities. Males die more frequently at birth. Males are relatively less developed at birth – and boys subsequently lag behind girls in reaching each developmental watershed. On every scale and measure possible, males are more infantile and as a result more vulnerable to disturbance. While all humans are paragons of youth, men are unequivocally the biggest babies of all.

THE BABY RACE

Imagine that you are a female primate – the species is not important – and you have spent nearly all your life alone on a remote island. All the other members of the population were killed by a giant tidal wave when you were very small. Desperate to find a mate, you decide that it is time to swim over to another island that is just visible on the horizon. After two days of swimming, you finally arrive, overcome by exhaustion and anxious excitement at the thought of finding others of the same species. Suddenly, you notice the most gorgeous male sauntering over from the shade of some palm trees. It is love at first sight and you are consumed by the feeling that this is the male with whom you want to spend the rest of your life. But hang on, you know nothing about him. Even if you and he mate, is he going to make the faithful and devoted partner that you came all this way to find? There is only one sure-fire way of telling: look at his testes. Oh no, they are huge – he is promiscuous – your worst nightmare has come true, and it is a long way home.

Testicles, or 'testes' as they more usually called by biologists, are terribly useful things. Hanging, as they normally do, pendulously from the rear end of an animal, they provide their male possessors with their only link with the future, in the form of sperm. They also provide good advertising material, as fabulously demonstrated by mandrills with their rear-end display of blue bottom and bright red scrotum – a feature that accurately replicates their front-end display of blue cheeks and bright red nose. To scientists too, they are immensely useful – and not only to male ones – since the testes size tells them how much sperm the males need to produce, and this itself is closely related to the mating system of the species. The rule is extremely simple, and is that males produce relatively more sperm in species where females mate with a number of males. In other words, it is not the number of females that a male mates with that affects the size of his testes – even males with huge harems have relatively small testes since they need produce only a tiny amount of sperm to fertilise each female – but rather the level of promiscuity of the female that is important. The more promiscuous the female, the more sperm a male has to produce to out-compete the other male's sperm inside the female's reproductive system. Male silverback gorillas, therefore, with their exclusive sexual access to the four or so females in their harem, have tiny testes weighing a mere 10g each. Male chim-panzees, on the other hand, weighing a quarter that of male gorillas but with their highly promiscuous mating system, have testes that weigh a massive 60g each.

Where then are humans on this scale? As it turns out, we are exactly where we would expect to be – somewhere in between, with human testes being neither large enough to indicate a high degree of promiscuity, nor small enough to indicate strictly faithful monogamy. Human testes weigh about 20g each. Or rather, most of them weight about 20g. To be precise, 104 Danish men were found to have testes that each weighed about 20g, although similar weights have been reported from studies of men from other European countries. As for the average testis size of Africans, the only source of information seems to be a report by the anatomist Adolph Schultz, published in 1932, in which it was stated that the average weight of the testes of three recently deceased black Americans, who we are told were 'unclaimed bodies of the anatomy department of the John Hopkins University', was 25g – that is, 25 per cent larger than those of Danish men. On the other hand, one hundred Chinese men from Hong-Kong had testes that each weighed about 10g. Even after taking into account their slightly smaller body size, Chinese men's testes were still found to be close to half the size of those of Danes.

So what is the point of dwelling on this – by all accounts – rather unattractive and ungainly part of the male anatomy? The answer is that testes introduce a second factor that influences the extent to which each of us is infantised – that is, what part of the world our ancestors came from. The reason is that testis size almost certainly provides a strong indication of the overall degree of infantising.

Despite a distinct reluctance among the scientific community to apply the general relationship between testis size and mating system to our own species, the small size of Chinese testes almost certainly tells us a great deal about their ancestors' sexual behaviour. According to the rule established by all other primates, the most likely reason why the Chinese possess relatively small testes is because they are descended from a long line of more faithful sexual partners. If the males have small testes, it is because the monogamous pair-bond is stronger and the female has very few 'extramarital' copulations. Of course, there will be the odd exception, and both sexes occasionally break the rule – as with the Emperor Fei-ti – but on average, more-or-less faithful monogamy seems to have been the modus operandi for the Chinese.

Furthermore, according to evidence that has already been given, monogamy is strongly linked to the infantising of our species. Before our ancestors regressed into a permanently immature state, promiscuity ruled and males and females mated with all and sundry. Then came female selection and a preference for infantile males who would transfer their dependence on their mother directly to their sexual partner. As our species became more infantile, so the bonds between males and females became stronger, and consequently monogamy took over as the primary mating system. It follows, therefore, that within our species, the more monogamous populations – such as the Chinese – are those that have been infantised to the greatest degree. In support of this, studies in America have found that people of Mongoloid origin have particularly stable marriages compared with the rest of the population. James Garbarino and Aaron Ebata from Pennsylvania State University have reported that those of North-East Asian descent living in the United States typically have significantly lower than average divorce rates and out-of-wedlock birth rates. One final point, before any women feel the urge to leap into bed with a tape measure to check on the size of their partner's testes, research has also found that within a particular race there is no relationship between testis size and the frequency of sexual activity.

Of course, the suggestion that the Chinese, or more broadly the Mongoloid people, are more infantised than the rest of our species, is entirely in keeping with evidence that has already been presented. One only has to cast one's gaze slightly wider than Chinese men's underpants, and take a look at the whole of their bodies, to appreciate how the Mongoloid body displays a large variety of extremely infantile characteristics. As discussed earlier, their relatively delicate skeletal bones, their hairless skin, flatter faces, bulbous heads, persistent inner eye-folds and many other anatomical features, confirm beyond any doubt that their bodies have undergone far more extreme infantising than in any other human race. Mongoloids also have the most extended developmental periods. For example, Chinese four-month-olds are considerably less active, and less vocal than European-American babies of the same age. A more protracted development may also explain the long life-span of many mongoloid people. And finally, according to some – including Philippe Rushton, professor of psychology at Ontario University – adults of Mongoloid origin have brains that grow to be significantly larger than those of other humans.

Behaviourally, too, there is overwhelming evidence that the Mongoloid people are at the furthest end of the infantile scale. In his book *Pink Samurai*, Nicholas Bornoff describes how, beneath their often steely exterior, the Japanese are highly mother-fixated and how the earliest poems demonstrate that this extends far back in history. The psychiatrist Takeo Doi similarly considers a childlike dependency to be one of the key components of the Japanese psyche. Transvestism and transsexualism, too, are more noticeable in the Far East, as well as in the Mongoloid Native North Americans, than anywhere else in the world, while Nicholas Bornhoff asserts that paedophilia in Japan is 'a quirk so widespread as to constitute a national trend'.

But perhaps the most impressive behavioural evidence for the extreme infantising of the Mongoloid people comes from the high incidence of homosexuality within many Far Eastern cultures. In his book *Passions of the Cut Sleeve*, historian Bret Hinsch describes how homosexuality in China has roots that go back at least as far as the Bronze Age:

> In many periods homosexuality was widely accepted and even respected, had its own formal history, and had a role in shaping Chinese political institutions, modifying social conventions, and spurring artistic creation... travellers from the West were quick to divide the world into a morally virtuous Occident superior to what they saw as the exotic decadence of the Orient ... Thus homosexuality became a focal point for division between China and the West.

The Emperor Ai, for example, who ruled China from 6 BC to AD 1, had a highly celebrated homosexual relationship with one of his subjects who was called Dong Xian. In fact, such was the emperor's infatuation with Dong Xian that he heaped on him, as well as every other member of Dong Xian's family, so many honours and privileges that the prime minister of the time publicly announced that the emperor's love of Dong Xian was plunging the country's affairs into chaos. In the ninth century, early visitors to China were astonished by the prevalence of homosexuality in this part of the world. Two Arab travellers of this time noted how the Chinese were addicted to sodomy and even performed it in their shrines. Likewise, the sixteenth-century chronicler Galeote Pereira reported, 'The greatest fault we do find [among the Chinese] is sodomy, a vice very common in the meaner sort, and nothing strange among the best.' Western travellers of every description wrote of how shocked they were by the ubiquity and depth of homosexuality within Chinese culture. The Dutch soldier Hans Putnams, who attacked the Fujian coast of China in the early seventeenth century, referred to men of this region as 'filthy pederasts', while Friar Gaspar de Cruz even went so far as to portray China as a new Sodom, despairing at the 'filthy abomination which is that they are so given to the accursed sin of unnatural vice, which is in no wise reproved among them'.

Homosexuality in China has, at various times, spanned all social classes and even included same-sex marriages. Shen Defu, a renowned literary figure of the Ming dynasty wrote in the seventeenth century of how two men could be married by adopting 'older brother' and 'younger brother' roles. A similar terminology was used in Japanese culture at the same time for two men who formed a 'troth of brotherly love'. Lesbians have also been allowed to get married. Within the 'Golden Orchid Associations' of southern China, a lesbian couple could choose to undergo a marriage ceremony in which one partner was designated as 'husband' and the other as 'wife'. These married lesbian couples could even adopt female children, who in turn could inherit family property from the couple's parents.

In Japan too, the tradition of homosexuality is extremely long. In fact, according to author and historian Tsuneo Watanabe, 'it is impossible to understand the traditional civilisation of Japan without taking it [homosexuality] into account'. As with the first Western visitors to China, those who first visited Japan were appalled at the widespread

acceptance of homosexuality as an integral part of society. After land-
ing in the Satsuma province of Japan on 15 August 1549, Francis
Xavier, the first Jesuit missionary to visit Japan, angrily wrote, 'There
are bonzes [Buddhist priests and monks] who love the sin abhorred
by nature; they admit it themselves; they never deny it ... nobody,
neither man nor woman, young nor old, regards this sin as abnormal
or abominable.' For many centuries before Xavier's arrival, homosexual relationships had long been the cause of incessant disputes within
Japanese high society, and were the subject of many historical writ-
ings. A fifteenth-century document, for instance, describes the case of
the 'Kakitsu affair', in which Yoshinori (1394–1441), the sixth shogun
who, having spent his early life in a monastery, surrounded himself
with young and beautiful male performers to satisfy his homosexual
passions. It is well recorded too that the ancient Samurai often
preferred to go into battle with one of their lovers at their side who
they considered was far more likely to protect them than somebody
who had no affection for them. As recently as the end of the nineteenth century, a European living in Tokyo wrote:

In peace as in war, the Japanese soldier marches arm in arm with
the friend with whom he is in an intimate relation. We can say, in
fact, that in the homosexual liaison too, the old samurai spirit
found exultant expression on the Manchu front [in the 1880–83
war between Japan and China], in a way that one would not have
seen before 1868. Many officers have told me of scenes where a
soldier in love with another had fought at the risk of his own life,
rushing willingly to the deadly spot. This is not simply due to the
warrior spirit and contempt for the death characteristic of the
Japanese soldier, but also to their passion for another soldier.

With the eclectic weight of anatomical, historical and cultural evidence pointing towards the conclusion that the Mongoloid people rest firmly at the most infantile end of the human spectrum, the time has now come to ask why this should be the case. Why should this particular group of humans that live, or at least originated, in the Far Eastern countries of the world have travelled so much further down the infantile route than all others? To start with, and stepping back slightly, it can be seen that rather than being isolated the Mongoloid people are instead at the end of a continuum. It is an extraordinary phenomenon that as one travels further away from the equator – which in the case of our ancestors meant north since this was the only way to get out of our species' cradle in Africa – so there is a tendency for people to become more infantile.

As soon as our ancestors moved northwards and started to inhabit the temperate parts of the world, the whole pattern of their existence would have changed. They would have experienced climates where plant food was seasonal and where meat became the only means of survival during the cold winter months. Unlike those living in the tropics, northern people had no choice but to become specialist big-game hunters – for mammoths, horses, and reindeer – and this required a far greater level of cooperation than had been needed before. What is more, during the cold winter months, females would have depended much more on hunted or scavenged meat, as plant food became increasingly hard to find.

Of all the temperate and cold areas that were inhabited by our ancestors, none would have presented more of these problems than the open grassland environment of the central Asian plateau where the Mongoloid people evolved. With the grasslands providing very little nutrition in the form of berries, nuts or root tubers, subsistence would have depended on hunting groups managing to bring down large mammals. In such an environment, one of the most effective ways to catch large mammals is to drive them into narrow ravines where they can be ambushed by others who are lying in wait. Alternatively, herds of plains animals could also be surrounded and driven over cliffs, into bogs or into the loops of rivers.

Either way, to survive in an environment such as the Asian plateau, early humans had no choice but to hunt in considerably larger groups than those living in warmer or more wooded areas, and then share this food with females and their children. From the females' point of view, whereas it might have been possible for them to fend for themselves and their children in warmer climes, on the Asian grasslands it was out of the question. To survive they needed a diligent sexual partner who could be relied upon to provide them with regular meat. With such strong pressure coming from females and the environment – for pair-bonding and hunting – it was virtually impossible for the Mongoloid's ancestors to avoid becoming exceedingly infantile, monogamous and cooperative.

In the light of this, it should come as no surprise to learn that the Chinese lived in larger and more complex societies far earlier in history than the rest of the world's cultures. Being more highly infantised, it is also to be expected that those of Mongoloid origin are naturally more gregarious and cooperative. It is also tempting to consider the possibility of a link between the degree of infantising and the political system that a country adopts – with more infantised groups tending towards more egalitarian, socialist political systems, while less infantised and more competitive members of our species are more inclined towards capitalist, free-market economies. In a recent interview, the Chinese scientist Xin Mao, voiced his opinion that 'The Chinese culture is quite different, and things are focused on the good of society, not the good of the individual'.

THUMPERS AND THINKERS

The third reason for variation in people's infantile qualities is simply to do with the advantages of being different. It is to do with what biologists call 'alternative strategies', which simply means there are different ways of achieving the same goal. The idea is that if the majority of the group is doing something one way, such as males fighting for females, then it is often better to pursue a slightly different strategy, such as covertly mating with females behind the other males' backs. Of course, as soon as too many individuals pursue the same 'alternative' way of doing something, then that starts to become the most common strategy, whereupon it pays to go back to the first strategy or find a completely new one altogether.

The natural world is replete with examples of alternative strategies. Within some species of frog, for instance, there are some 'caller' males who croak loudly to attract a mate, while other 'satellite' males silently lurk around hoping to mate with females who have been attracted to a caller. At first sight, it would seem as though all males should croak – until, that is, one learns that croaking not only attracts females but also predators such as bats and snapping turtles. In this instance, so long as there is roughly the right proportion of calling and satellite males, both strategies will have roughly equal chances of success. A similar situation occurs in field crickets, in which some males call – and thereby attract females and parasitic flies – while other males hang around hoping to catch the occasional female while avoiding most of the killer flies. From sticklebacks and beetles to digger wasps and birds, alternative strategies have been recorded from a wide variety of different species.

As applied to the infantising of the human species, precisely the same situation exists, and it can pay huge dividends to adopt a different strategy from the majority. If the majority of the population is behaving, say, in a macho and aggressive way, then evolution is likely to favour those who behave more cooperatively and gang up to achieve what they want. On the other hand, if the majority are meekly cooperating, then it might be better to be more dominant and exploit the weakness of the majority. The result is that there is constantly a balance of humans of varying infantile states, with each type carving out a niche for itself and thereby avoiding unnecessary competition with those with different skills and qualities. Of course, in different environments there will be a slightly different balance, with some infantile states being of more use than others. But nevertheless, because of the complex nature of our societies, there will always be immense pressure for division of labour and for people with particular talents to contribute in their own specialised way. In the same way that a society full of generals does not make sense, neither too does a society full of artists. It is all a matter of balance, and for that a variety of infantile conditions is required.

Even in extreme situations, variety is still highly valuable. Imagine, for example, the composition of an army. While at first sight it might seem that only one 'aggressive' type of human is needed, there are in fact a huge variety of qualities that are required to create an efficient fighting force – from leaders, to willing followers, inventors of weapons, makers of weapons, doctors, priests, and so on. We are far too talented and complex a species to be able to excel in all fields, and so we vary – with different individuals possessing different skills that they pass on to their offspring. What is more, over many generations, families become polarised as they specialise in different tasks – as skilled parents pass on their genes and knowledge to their offspring – and this inevitably means that marked differences appear in the amount of infantile characteristics that different families retain into adulthood. While this is not to say that each of us is rigidly destined to a particular 'place' in society, it is recognising the fact that society is composed of people with widely varying personalities and talents, and that different people excel at different tasks.

THE PERPETUAL PUPPY

So there we have it – our destiny lies precisely encoded in our genes. Depending on whether we are male or female, what part of the world our parents came from, and what our ancestors' particular social strategy was, we are born with genes that make us more or less infantile, and that is the end of the matter. Or not. Despite muddying the water, this is most definitely not the whole story. Consider the following comments made in a scientific article on animal domestication made by Edward Price from the University of California's Department of Animal Science:

Captive young animals may be conditioned to retain their juvenile behaviors. Offering positive rewards for juvenile behavior (e.g., care-soliciting, playfulness, submissiveness to humans) may, in effect, retard the development of more independent adult activities, or alternatively, may mask their expression.

Rearing animals in physical isolation from older, socially dominating [members of the same species] may preclude the development of normal adult-like patterns of agonistic behaviors and may result in the retention of many juvenile social behaviors.

Genes do not work in isolation. While genes affect the body from the inside, the environment affects the body from the outside. For example, although our genes determine the rough dimensions of different muscles in our body, the final dimensions of each of these muscles depends to a large extent on our environment and the types of activity that we undertake. Take two identical cloned humans, and put one in a rowing boat for four hours a day, while the other one watches television, and despite having identical genes, they will have very different-shaped bodies. And so the same is true for our behaviour. Although our personalities have a definite genetic element – with certain personality traits, for example, running in families – our behaviour is enormously affected by our previous experiences and our environment.

The same applies to our rate of development. While there is no doubt that our species' infantising has involved changes occurring to the genes that control the timing of our various developmental phases, the environment also has an important part to play. It is common to hear of children who have had to 'grow up' extremely fast because of the predicament that they found themselves in – children who, often tragically, have no choice but to expedite their development and exchange playful dependency for mature responsibility. In some environments, infantile behaviours are clearly not simply inappropriate or inadequate, they are lethal. For instance, childlike inquisitiveness and dependency are neither as appropriate nor as possible in war-torn Sarajevo or Rwanda, as in the wealthy suburbs of London, Paris or San Francisco. It is an essential part of our biology that the amount of infantile behaviours that we retain into adulthood – just like other aspects of our personality – is affected by our environment as well as our genes.

'Alas! That such affected tricks should flourish in a child of six!' was how Hilaire Belloc expressed the prodigiously manipulative talents of children. Infantile behaviour is, above all else, adaptive. My ten-year-old son still cries on the odd occasion when he 'wipes out' on his skateboard. It is not that crying repairs his body, it is simply that it significantly raises the chances of him receiving more attention. The fact that we occasionally cry alone is simply because this particular behaviour is hard-wired into us and does not take into account the presence or not of an audience. Children behave the way they do, and not like adults, because countless generations of children who have behaved this way have done better for themselves than children who behaved differently. Infantile behaviour, in other words, adapts a child perfectly to its particular environment. Later on in life, when their environment changes, and when they are surrounded by people with different agendas to those of children, then their behaviour changes accordingly. They discard their childish behaviours for another set that will maximise their success in a more mature environment. Swapping, exchanging and replacing behaviours as we get older is what we call 'growing up'.

But what if the environment does not change? What would we expect the growing individual to do then? Would we expect it to change its perfectly successful set of behaviours for another that gives it fewer rewards? The answer, going by observations on our own species as well as those of other animals, is no. It seems that certain environmental conditions cause us to become even more infantile than we are already.

This provides an explanation for why it appears to be the case that the more affluent a society becomes, the more infantile its members remain. Put another way, the more a state takes over responsibility for every aspect of its inhabitants' lives, the less need there is for each citizen to develop mature characteristics. As long as the people from a very early age are provided with all they need – in terms of food, housing, healthcare and entertainment – then there is no need for them to develop more mature, competitive and selfishly aggressive behaviours. By offering positive rewards for being childlike and submissive, there is every likelihood that such behaviours will remain for as long as the rewards continue. Indeed, the whole infantile scenario can be strengthened by not only providing people with all that they need, without the need to 'fight' for these, but also by threatening to withhold these comforts if they should behave in a mature, intolerant and antisocial way towards each other. Such punishments can either be immediate, such as incarceration in prison, or longer term, such as 'Hell' and other diabolic places. Even better is to combine all these things with persistent reminders of the constant threat from 'outside' and the horrendous consequences if support is withdrawn. As Noam Chomsky famously said: 'The more you can increase fear of drugs and crime, welfare mothers, immigrants and aliens, the more you control all the people.'

An even more effective way of creating dependency was 'discovered' by the American psychologist Theodore Reik over fifty years ago when he noticed that people can become highly dependent on those who dominate them through a combination of extreme kindness and cruelty. A fascinating experiment conducted on dogs in 1955 illustrates this phenomenon perfectly. In this experiment, three sets of puppies were raised under very different conditions. Puppies in the first group were consistently provided with love and kindness. Those in the second group were consistently punished every time they dared to approach the scientist. And finally, those in the third group were treated inconsistently – sometimes being cuddled and petted, and sometimes, for no reason at all, being punished. The result was startling. The most affectionate, dependent and submissive of all the puppies were those that were provided with an inconsistent mixture of punishment and kindness.

Treat people like children, reward them for doing so, occasionally threaten them, and there is every reason to believe that they will retain their childlike behaviours into adulthood. Of course, whether this is a good or bad thing is quite another matter. On the one hand, it is undoubtedly highly manipulative. On the other hand, it is potentially a recipe for a society that is peaceful, playful, creative and extraordinarily dependent on the government for every aspect of its well-being and safety. Create a nanny state and the result is 'kindergarten mentalities'.

Identical scenarios are frequently witnessed in homes and zoos across the world, where juvenile behaviour is extended into adulthood as a result of the animals being provided with all that they need without having to fight for it. Among domestic dogs, as has already been mentioned, mollycoddling during their early stages of life can result in the 'perpetual puppy syndrome'. As a result of becoming overly dependent on their owners, such dogs are frequently highly sensitive to disturbance and can exhibit a variety of neuroses ranging from compulsive eating to attention-seeking paralysis. Animals in captivity can display similar behaviours. Squirrel monkeys, for example, have been found to retain essentially infantile or juvenile calls into adulthood when they are raised in captivity because they have no need to develop the mature calls that are associated more with competition. Furthermore, in relation to the higher levels of curiosity seen in captive primates, Martin Moynihan suggested that 'juvenile levels of investigatory behaviour persist in the captive adult through being maintained in a dependent, or juvenile, role by keepers, who act as substitute parents'.

The same phenomenon has been observed in birds. For example, if starlings are hand-fed well beyond the chick stage, they continue to show a chick-like 'gape' response – that is, they open their beak as wide as possible waiting for food to be dropped into it – despite being quite capable of pecking for food themselves. Similarly, psychologists Mertice Clark and Bennet Galef found that captive gerbils fail to develop adult exploratory behaviour if they are provided with plenty of safe dark tunnels, and concluded that: 'the effect of shelter-rearing on the development of exploratory behavior is interpreted as demonstrating an environmentally induced maintenance of an adaptive juvenile pattern of behavior'. On the other hand, Norwegian rats kept in cosseted laboratory conditions retain an infantile tolerance of foreign males rather than developing the aggressive behaviour of their mature wild relatives. The picture is the same across the animal kingdom – if juvenile behaviour works, why change it? Or, in the words of a traditional saying, 'if it ain't broke, don't fix it'.

So what of our own species? Is it also the case that pampering and overprotection cause humans to remain even more infantile – or 'super-infantile' – throughout adulthood? Evidence all around us certainly points in this direction. From the foppish, playful rich – unburdened by everyday competition, stresses and strains – to the vibrant city environments where adults of every age can indulge their playful predilections in a myriad different ways, there is good reason to believe that affluence may well go hand in hand with a more infantile character.

The situation bears a distinct resemblance to the studies mentioned above, in which animals that are automatically provided with everything they need have no reason to develop mature assertive behaviours. However, what animal experiments also show is that, once lost, infantile behaviours are unlikely to reappear – regardless of the environment later on. For example, once a dog develops 'wild' mature characteristics, it is very difficult to tame later on, even if it is subsequently provided with everything it needs. What this suggests is that the upbringing of humans is central to the process of 'super-infantising', and that for an individual to remain in an extremely juvenile condition throughout adulthood, they must have been raised under conditions that encourage the retention of childlike behaviour, and not simply exposed to such conditions later in life. Consequently, there will be something of a time lag between the introduction of all-embracing welfare systems and the appearance of the infantile behaviour throughout the whole population. Clearly, infantile compliance and dependency cannot become pervasive until a whole generation has been brought up under 'nanny state' conditions.

That the removal of stress encourages the persistence of infantile behaviours is extremely difficult to investigate, given the number of other cultural factors that complicate matters. However, there is evidence that since Western countries have become more welfare-oriented, so people living in these societies have had their infantile phase hugely extended. A report published in 2001, for example, found that young people living in the USA and England now consider themselves to reach adulthood around the age of about thirty-five. Only at this age did many of their attitudes, tastes and aspirations change dramatically from those held since adolescence. Peter Francese, founder of *American Demographics* magazine, similarly found that thirty-five was the modern age of maturity. 'At 25, most graduates are economically useless, almost unemployable, changing jobs at a whim; how can we call them adult? But at 35 they are getting serious, talking about careers rather than jobs and, most importantly, having children.' To some, perhaps especially the wealthy, even thirty-five is too young to be considered grown up. George W. Bush has dismissed many of the things that he did before his fortieth birthday, including drink-driving, as merely 'youthful indiscretions'.

There is one widespread behaviour that provides strong evidence that particular environments can increase the chances of certain individuals retaining infantile behaviour patterns into adulthood. It concerns one of the most persuasive indicators of infantising – homosexuality. For those at the infantile end of the human scale – who teeter on the genetic brink of remaining within the single-sex stage of development – it can be predicted that a more protective environment will increase the chances of their infantile behaviour, including same-sex attachment, being retained into adulthood. 'Environment', of course, does not simply mean the way that a child is raised by his or her parents. As Judith Rich Harris demonstrates in her book *The Nurture Assumption*, parents have astonishingly little effect on a child's personal development – including their attitudes, accent and vocabulary – and life outside the home has far more influence than was previously thought. The environment of a child therefore covers everything that he or she encounters, including school, advertising, peer relationships and sport, in addition to parents and a myriad other influences.

There is indeed evidence that homosexuality in the Western world is significantly more prevalent in the most playful and nurturing environments – that is, in cities. In America, more than 9 per cent of men in the nation's twelve largest cities identify themselves as gay, compared with 3 to 4 per cent of men living in the suburbs, and about 1 per cent of men in rural areas. As for the criticism that homosexuals move into cities because of prejudice elsewhere, this seems not to be the case. The following quote is from *Sex in America: a definitive study*, which followed an extensive research programme that was conducted in the early 1990s:

> It is not just that homosexuals tend to move to large cities from the small towns and rural areas where they grew up. In our survey, we asked people where they were living when they were fourteen years old, and, with those data, we discovered that people who were raised in large cities were more likely to be homosexual than people who were raised in suburbs, towns or the countryside. This relationship also showed up in the General Social Survey, an independent national sample.

Because cities by their very nature require people to live in far higher concentrations than elsewhere, there is even more need for those living there not to mature too far. This is especially important in preventing the emergence of mature levels of aggression and intolerance towards strangers who, of course, those living in cities constantly encounter. And just like the captive Norwegian rats referred to earlier, the best way of achieving such a society is by treating cities' inhabitants like dependent children. The more that humans have responsibility for looking after themselves taken away from them, the more childish and amenable they become. As with hamsters, starlings and squirrel monkeys, it is all a matter of rewarding people for infantile behaviour and thus giving them no reason to change. The more human societies reward people for remaining childlike, the more cooperative, playful, compliant, and incidentally homosexual, they are likely to become. It also, incidentally, suggests that societies in which boys are indulged more than girls are further contributing towards the higher incidence of homosexuality in men than women

So just how big a part does the environment play in determining how infantile each of us remains during adulthood? As mentioned previously, since virtually every aspect of our biology depends on the interaction between our genes and our environment, including behaviour such as docility and aggression, there is no reason to believe that the extension of infantile features into adulthood should be any different. Even though there is undoubtedly a strong genetic influence on the degree to which we fail to mature – as illustrated by the fact that homosexuality runs in families – our environment will inevitably affect the extent to which this actually happens. Moreover, if scientists such as Simon Conway Morris of Cambridge University are correct in thinking that our species has remained genetically much the same for the past 50,000 to 100,000 years, then our environment has been responsible for any shifts in behaviour, including becoming more infantile, that may have occurred since then. As to whether we have become more infantile within the past few hundred or few thousand years, this is extremely difficult to tell for sure. However, going by the childlike dependency of many of those who live in industrialised countries, as well as the high incidence of homosexuality in cities, it certainly seems to be the case.

To a vast extent, life's rich tapestry revolves around people display-ing more or less infantile behaviours. From the belligerent but highly protective mature matriarch or patriarch, to the diligent and corrigi-ble office worker, or the infantile 'head-in-the-clouds' artist, so much of life's variety is created by individuals retaining a variety of differ-ent immature behaviours into adulthood. Francis Galton's axiom that life is a complex mixture of 'nature and nurture' is never more appropriate than when applied to the extent that each of us remains infantile throughout our lives. While our genes provide our overall level of infantising, exactly how we turn out depends on the precise conditions that we encounter as we grow up.

chapter ten

the four
human types

We have at last reached the final leg of the journey that began over six million years ago when our species began its journey down the road to permanent babyhood. The result is a species that displays a dazzling variety of infantile features, both in its appearance and behaviour. Moreover like all things biological, there is variation – indeed, the world would be a very drab place without it – and it is the details of the way that humans differ that forms the subject of this chapter. To fully appreciate the central effect that infantising has had on human evolution, a brief recap. Initially, as Africa became increasingly drier, our regression was driven by the massive benefits of living in large infantile super-gangs. Unlike their mature somewhat anti-social predecessors, our first ground-living ancestors could terrorise the predators through sheer force of numbers. Incidentally, our ancestors' bodies became more infantile – they had smaller teeth, flatter faces, and they started tottering around on two legs. Then, around two million years ago, as the environment took another turn for the worse, the second major wave of infantising occurred. To increase male devotion and bonding, females began to selectively mate only with those males who retained a particularly infantile need for mothering throughout their lives, and who would therefore transfer their dependence from their natural mother to their sexual partner. Monogamy, permanent maternal swollen breasts, and a substantially more infantile anatomy – including our ridiculously swollen and baby-like heads and brains – were the result. And finally, sometime between 125,000 and 50,000 years ago, the chance emergence of a mental state called 'consciousness' irreversibly converted our ancestors from animals to humans, and with this came art, language, technology and religion. For the first time our ancestors could 'daydream' – look at one thing and think about something quite unrelated. Unshackled from reality and the present, they could chip away at a stone tool while thinking about sex – or have sex while thinking about a new tool.

Glimpses of what human society would have been like if we had not become so extremely infantised are occasionally seen in zoos where other primates are placed in close confinement. One example occurred in London Zoo at the beginning of the last century, when its director at the time made the decision to house a large number of male and female Hamadryas baboons in a single enclosure. The result can only be described as 'rivers of blood', as males and females died in horrific numbers as they fought to the death for high-rank and access to sexual partners. Similar scenes of intolerance and raw aggression occur when members of our closest relative, the chimpanzee, are housed in mixed-sex groups in zoos – with males in particular using their massive strength to assert their dominance and status, and often fighting to the death to maintain their rank. Had our own species not been infantised to the degree that it has, there is no doubt whatsoever that identical scenes would constantly be witnessed in our own species. However, as it is – and as mentioned earlier – what is absolutely astonishing about cities such as New York, Nairobi or London is not the occasional fight, mugging or rape, but the total absence of any aggression whatsoever during the other hundreds of thousands of interactions that occur each day. The greater the need for people to live together in large numbers, then the more necessary it is that society becomes more infantile. Like meek, subservient children, the masses need to do what they are told – accepting their lot with infantile obedience and subservience.

But clearly human societies are not composed of identically infantised individuals. All people are definitely not all equally infantile, cooperative and sociable. Within the infantile constraints imposed by females – who have, for thousands of generations, selected dependent males to be their sexual partners – some people are decidedly pushy and aggressive, while others are so infantile that they would hardly dare say boo to a goose. For reasons given earlier, there are a number of factors that affect how infantile we remain throughout adulthood, some being genetic, while others are to do with our upbringing and environment. So how does all this relate to our own characters and everyday behaviours? The answer is that, while most of us float somewhere around mid-way between the mature and infantile extremes, there will inevitably be those who veer strongly towards one extreme or the other. Although we are all infantile at heart, some of us are considerably more infantile than others. It is something that we are innately aware of, yet few appreciate.

From the moment that we first lay eyes on a stranger, we start to construct a profile of his or her character. We extrapolate from one or two items of information that we glean about a person to build up a complete 'predicted' profile. On the basis of a remarkably few pieces of information, we often feel confident about predicting a whole gamut of other aspects of a person's lifestyle and attitudes. Indeed, it is essential that we do so in order to adjust our behaviour accordingly and get the most out of our interactions or relationships with other people. Unknowingly, we are constantly pigeon-holing everybody around us into certain categories. Once categorised, we then begin to make various predictions about how they are going to behave in various different situations. We refer to our stored data on this 'type' of person to predict a huge variety of different aspects of their character – their likes, dislikes, attitudes, political and sexual leanings etc. There will of course be minor variations – details that differ from one person to the next depending on their particular experiences – but on the whole it works with remarkable success. But why does it work so well? The reason is, as many psychologists have argued, that within our species, people tend to display well-defined collections of personality characteristics.

Hans Jürgen Eysenck, a man who fled Nazi Germany in 1934 and later became Professor of Psychology at the University of London, is perhaps the most famous early proponent of what are called 'trait theories of personality'. Using a battery of questionnaires, Eysenck spent many years looking for aspects of personality that typically appear together, and out of this created a scheme that showed how people's personalities are made up of clusters of linked characteristics. Despite having a rather too rigid adherence to inherited rather than acquired features, Eysenck's basic model of personality has neverthe-less proved fairly robust in many experimental situations. Recently, other psychologists have developed further schemes and refined the major defining elements of personality. The most important of these is the 'Big Five' model, in which people can be defined according to five basic qualities that they may or may not possess – whether they are extrovert and outgoing, agreeable and sympathetic, efficient and organised, anxious and unstable, and artistic and imaginative. What is more, a growing body of research suggests that the Big Five model provides a descent description of personality not only in adulthood but also in late childhood and early adolescence, and that it applies across a wide range of different cultures. But this is far from the end of the matter. As reliable as psychological tests such as these may be, they still fail to address the basic problem of what causes different people to acquire different personalities. The answer, I suggest, is that people's personalities differ largely as a result of them being infantised to differing degrees.

In any population, the majority of men and women will remain close to the average condition for our species, and that goes as much for the degree to which they retain childlike qualities into adulthood as any other feature. This majority will fail to develop anything like the mature levels of aggression and intolerance of our distant ancestors, and remain cooperative and playful for the whole of their lives. But within this broad group, there is considerable variation, and different individuals will show distinctly different characteristics depending on which side of the average they fall. Those on the mature side will be less playful and novelty-seeking, but nevertheless be cooperative and fit in well to well-structured work environments. These are the 'Bureautypes'. Those on the infantile side will be more playful and creative, more rebellious, and find it harder to fit in with rigid work hierarchies. These are the 'Neotypes'. Together, the Bureautypes and

Neotypes form the core of human society. But there are also the extremes – both of which have their place in society, but which form the minority. At the mature extreme are those who retain so few infantile characteristics that they develop a strong desire to achieve high status by whatever means . These are the 'Alphatypes'. At the opposite extreme are those who have been so infantised that they not only remain exceptionally childlike, playful and insecure throughout their lives, they also fail to develop a sexual attraction towards members of the opposite sex. These are the 'Ultratypes'.

By and large, most of us will fit into one of the four categories. Depending on our genes and our environment – especially that of our childhood – we will possess a collection of more or less infantile behaviours. Of course, there will be some who find that they rest between two adjacent categories, and show features of both. There are also some behaviours that appear the same, but have two distinct origins – such as territoriality that may either reflect infantile insecurity or a mature desire for dominance, and promiscuity that can arise through infantile playfulness or as a droit de seigneur expression of dominance.

THE ALPHATYPE

The Alphatype is the 'top-dog' or, to apply the metaphor fairly to both sexes, the 'top-bitch'. He or she is the type of human who retains the fewest number of infantile behaviours into adulthood, and most closely resembles our mature primate ancestor. If you know an Alphatype then you will think of them as being ruthless, determined and particularly concerned about their image as a strong leader. They hate to be shown up as being weak in any way, and are extremely intolerant of disobedience. At work, the Alphatype is bossy, ambitious, and focussed on their career, while at home they are domi-neering and firm that everything has to be done their way. The powerful matriarch or patriarch. Having diverged the least from the behaviour typical of other mature primates, the Alphatype is the most aggressive as well as most dominance-seeking member of his or her family, social group, or profession. To the mature primate, status is first and foremost, and play for the sake of it simply does not come into their behavioural repertoire.

Groups that consist entirely of Alphatypes are extremely similar to those of other mature primates inasmuch as they can be extremely aggressive for their size, while being fraught by divisive battles for supremacy. Consequently, all-Alphatype groups are dogged by a high degree of friction and in-fighting, and as a result they disintegrate well before they reach any significant size. The best way for Alphatypes to form large powerful groups – on the battlefield, in business or on the sports' field – is to team up with Bureautypes who, while they retain a certain degree of competitive drive, are more easily led and less determined to be top-dog. Team sports are, unsurprisingly, attractive to Alpha-types as a way of displaying their high-status and dominance. Alphatypes want to join competitive rather than creative groups, and this is reflected in every aspect of their lives including, for example, the types of pets that they own. From German Shepherds, to Dobermans and Rottweilers, the dogs owned and admired by Alphatypes reflect their self-image of being aggressive, assertive and dominant. As an extension of their owners' personalities, such dogs are likely to be encouraged to impress their strength and status on others, especially when it comes to fighting other dogs.

Of all the members of society, Alphatypes are the most likely to resort to brute force to solve a problem rather than taking time to contemplate how to solve it by other more manipulative and coercive means. As a result they certainly do not always form the best leaders. Whatever else, Alphatypes are not fearful of, nor do they avoid, conflict – if male, they are the 'Rambo' figures – be it in the battlefield or the boardroom. In evolutionary strategy terms, they are 'hawks' not 'doves'. Whereas doves try to settle disputes by signalling and retreat if attacked, hawks fight until they win or sustain enough injury to force them to retire. They will be the most keen to escalate hostility to establish, once and for all, who is at the top of the pecking order – regardless of whether this is in a bar brawl or a boardroom.

In being the least infantised members of our species, Alphatypes are also the least playful and inventive members of society. They are not interested in novelty, or artistic pursuits, nor do they get bored as easily as the more infantised types. Of all the human types, Alphatypes are the most goal-oriented and focussed. As applied to their everyday lives, this results in the Alphatype being less concerned about variation for its own sake, for example in relation to food or sex. In having a distinctly functional approach towards matters, they would certainly not be described as 'fun-loving'. Instead, Alphatypes are the world's greatest risk-takers. Since their goal is not to play but achieve dominance, they are willing to go the extra mile to achieve what they so desperately want even if this can mean taking great risks and losing everything in the process. As military, political or business leaders, this means that Alphatypes play with high stakes – an extreme strategy that tends to result in total annihilation of the opposition or themselves.

Being mature and independent, Alphatypes are unlikely to turn to religion or to their sexual partner for support. Their wives or husbands provide them with assistance in their conquest for power, as do their friends or colleagues. They do not provide a shoulder to cry on. For the Alphatype male, his wife is far less of a mother-figure than for other types of male. Life for the Alphatype is a battleground and his or her family is very much a part of the overall strategy to achieve domination. Insurrection in the family, as at work, is not tolerated. Being politically and status motivated, as long as their friends or family remain loyal then they return the favour. But they are not emotionally mawkish – they can swiftly drop their closest allies when their trust is betrayed or an alternative course of action offers far greater rewards. Because of their low emotional dependence on their sexual partner, and their high competitive nature, Alphatypes have a strong promiscuous drive. It is not that they seek extra-marital affairs in an attempt to satisfy an unfulfilled need for love and emotional support, but rather to confirm their feeling of a superiority over other males or females. To Alphatypes, lovers are the sexual equivalent of military honours or battle scars. In this, as with many other aspect of their behaviour, Alphatypes are displaying a very ancient feature of our species that predominated for many millions of years – that is, until females largely abolished it by choosing to mate only with the most infantile males who remained emotionally dependent on them as mother-figures for life.

THE BUREAUTYPE

With a slight increase in the degree of infantising comes the perfectly cooperative human being. While the Bureautype has not lost all the aggression and desire for status of the Alphatype, he or she is a considerably more cooperative animal. Instead of constantly wanting to be at the top of the group, as does the Alphatype, the Bureautype forms far more egalitarian groups that are largely lacking infighting and within-group aggression. Whereas Alphatypes are constantly squabbling for a higher position in the pecking order, Bureautypes are far more comfortable working together with a less-competitive hierarchy, and relying on others to take the lead as well as the risks. The Bureautype's greater immature insecurities means that they form cohesive groups, as each individual relies on the others for emotional as well as physical support. Although they are reasonably adventurous, they are also more uncertain of being left alone and exposed than Alphatypes. The Bureautype is far happier than the Alphatype being in close proximity with others and welcomes support from the rest of the team, allowing the formation of far larger groups. In lacking an excess of both aggression and competitiveness, they form the perfect army or 'flock', often being told what to do by Alphatype leaders.

To many people, our species seems highly aggressive because of the size and behaviour of armies throughout the world. This, however, is totally wrong. As mentioned earlier, the formation of large armies, just like living in large numbers in cities, is only possible because of an extreme lack of aggression. Take a moment to compare our own behaviour with that of our closest relatives, the chimpanzees. In this species, such are the levels of competition and aggression that groups any larger than 5–10 break down through tension, anarchy and competitive aggression. In contrast, humans are almost unbelievably peaceable in being able to cooperate with tens of thousands of other individuals to form armies in the first place. We simply would not be able to do so unless we had incredibly low levels of aggression within our species as a whole. War as we know it, involving vast hoards of humans – primarily Bureautypes – can be said to exist solely because we are so incredibly unaggressive towards members of our own group. Without this feature, we would revert to intense squabbling at an individual level. Although Bureautypes form the bulk of armies, they are not highly aggressive. Within their own group, it would be extremely rare for them to fight to point of death. However, during war, a combination of their intense loyalty and insecurity causes them to kill simply to protect other members of their hugely-extended 'family'. Killing, after all, can result as much from extreme fear as extreme hostility.

Bureautypes are more imaginative than Alphatypes, having retained more of their childlike playful and exploratory qualities into adulthood. The brawn of the Alphatype is more than compensated for by the brain of the Bureautypes. They are just the right mixture of ambition and creativity – not too aggressive and yet not so absorbed in their own 'play' as to be useless as a team member. Bureautypes are therefore good in organisational positions where they need to cooperate with each other and work towards a collective goal. It is the Bureautype that would have been the most favoured in the early stages of our evolution, when working in large cohesive groups was the only way that our earliest ground-living ancestors could withstand the horrendously high levels of predation. For them, cooperation rather than constantly squabbling to reach top-dog position, enabled the formation of the highly-efficient super-gangs. Because of their great insecurity, Bureautypes are more likely to seek guidance and leadership than Alphatypes, either from other people – most frequently Alphatypes – or in the form of spiritual leaders through religion.

It is this same combination of reduced ambition for high-status, and an increased infantile dependence on their sexual partner, that makes Bureautypes considerably more faithful sexual partners than

Alphatypes. Compared with Alphatypes, they have retained much more of a need for a mother- or father-figure into adulthood, with their sexual partner providing both sex and parental security. As with their ability to work well in a business environment, so Bureautypes are also ideally suited to family situations in which they have well-defined roles and responsibilities. Unlike Alphatypes, Bureautypes are not overly concerned about their status in the family situation, and unlike the more immature types of human, they are not excessively driven to indulge their own fantasies rather than look after other members of their family. Although they are not as fun-loving as Neotypes, Bureautypes make the most reliable parents and are extremely organised and level-headed about family issues.

THE NEOTYPE

As yet more immaturity is introduced into the adult human personality, so the next type of human emerges. Behaving in a way that is increasingly childlike, the Neotype always strives for novelty in his or her life. Neotypes fail to develop the vast majority of the aggressiveness that accompanies maturity, but instead remain immaturely insecure, playful and inquisitive. Instead of fighting for what they want, as would Alphatypes or to a lesser extent Bureautypes, the Neotype relies far more on childlike manipulation and coercion. There is, however, considerably more variation within this category than with the more mature categories, and this is to do with the greater breadth of personalities of younger individuals. While experience and confidence result in mature individuals occupying a fairly narrow band in terms of their range of behaviour, younger individuals are far less conformist and react in much more extreme ways to different situations. In large social gatherings, for example, children range from being extremely timid and 'clingy' to their parents, to being highly gregarious, self-confident and brash. So too for Neotypes who, because of their greater retention of youthful characteristics into adulthood, also possess a far wider range of personalities. While most Neotypes retain the exuberance and self-confidence of immaturity into adulthood, some remain insecure and timid. Other immature characteristics that Neotypes frequently retain into adulthood are petulance, moodiness, being easily influenced, and always wanting to be the centre of attention. Such extremes of behaviour are far less common among the more mature types of human.

The Neotype is most obvious as the fun-loving friend who is eternally playful and boisterous. Their out-going and alacrious personalities are reflected in every aspect of the Neotypes' lives, from the large number of friends that they have, to the pets that they own. Whereas Alphatypes favour aggressive fighting breeds of dogs, Neotypes prefer more playful breeds that display affection rather aggression and behave as play-mates rather than members of a fighting gang. They are also far more likely to own cats than Alphatypes because of the less threatening qualities of this animal companion. Above all else, Neotypes retain a strong desire to seek variety, a vital part of infant behaviour when it comes to learning about life and exploiting the environment to the full. As such, they have a somewhat short attention span because they quickly get bored and want to move on to find a novel form of stimulation. This does, however, mean that Neotypes are extremely creative and very imaginative members of society. Throughout their lives Neotypes remain highly imaginative and, when combined with the experience that comes from many years of life, this means that they are often the originators of revolutionary inventions, be they artistic or scientific. While often seeming to lack a firm grasp of reality and functional application, Neotypes can have an extraordinary ability to play with ideas in their mind and come up with highly original thoughts. Like so many other aspects of the Neotype's life, their creative talents very much arise through their childlike rebellious qualities.

As a result of their infantile nature, Neotypes often find it difficult to form long-lasting relationships. Unlike Bureautypes and Alphatypes, Neotypes remain more in the 'testing-the-water' or 'rehearsal' stage of sexual and social development, and find it difficult to face the thought of settling down to start a family. They are less confident about themselves, their own identity, and also what best suits them. When they do form 'permanent' relationships, Neotypes find it more difficult to remain faithful since they are constantly yearning for novelty in their sexual relationships as with other aspects of their life. Unlike Alphatype promiscuity, however – which is based on status – Neotype promiscuity is far more a matter of playfulness and rebellion.

Yet despite this, Neotypes make extremely devoted, loving and dependent sexual partners. Being infantile and insecure, they very much rely on their partner to act as a pseudo-parent and to provide them with emotional support – far more so than Alphatypes or Bureautypes. From the moment that they leave their parents' home, Neotypes crave mother- or father-substitutes on which to heap their emotional dependence, and their relationships are characterised as being extremely demanding both in terms of sex and emotional support. While Neotypes can be extremely passionate – this being an emotion which, after all, is derived from mother-love – they can also be very petulant, and if they are not receiving sufficient 'parental' love and attention from their sexual partner then they are likely to look elsewhere.

Neotypes are therefore provided with something of a dilemma. On the one hand, they possess a childlike dependence on their partner for emotional support, while on the other hand they crave variety and are constantly yearning for sexual novelty. They are like children who are torn between the comfort provided by their mother or father, and going off on their own to look for new exciting experiences. Clearly, one solution is to indulge in imaginary promiscuity, and Neotypes are highly likely to experiment with sex and incorporate fantasies into their sex-lives. Sexual fantasy has the potential to provide Neotypes with the best of both worlds – it allows them to experience sexual novelty, while simultaneously remaining in a relationship that provides constant emotional support. For this reason, and because of its origins in infantile dependence, subordination during sex, and ultimately masochism, is more likely to appeal to Neotypes than Alphatypes or Bureautypes.

Because of their infantile qualities, being married to a Neotype can be an emotional roller-coaster since it involves having to put up with the emotional insecurity as well as the occasional outbursts of emotion that Neotypes are likely to have. The Neotype is somewhat torn between desperately needing affection and stability from their sexual partner, while at the same time wanting to establish their own independence. Like children, Neotypes have considerable mood-swings – one moment they are elated and excited, the next they are in a sulk about something that has been denied to them. A positive side, however, of this rather turbulent character is that the Neotype is unlikely to harbour resentment for very long, and soon forgets any animosity after a disagreement or argument. Ultimately, their emotional dependence on their partner means that they need a cuddle more than a cold-shoulder, and often the best way to break the ice is to start to play with them. Like rebellious children, Neotypes simultaneously seek freedom to do whatever they want, while retaining a strong reliance on a parental figure for emotional support.

Neotypes also have a somewhat contradictory relationship with their children. While Neotypes can make wonderful playmates for their children because of the imaginative playfulness that they retain into adulthood, they never quite feel that they are ready to accept the responsibility for raising children. They consider themselves too young to have the burden of children and want to keep on enjoying themselves without any responsibility to others. To make matters worse, because Neotypes have such an emotional dependence on their sexual partners, children can be envisaged as an irksome source of competition for the partner's attention. Before the child arrived, the Neotype was able to play with his or her sexual partner and receive 100 per cent of their attention, whereas after a child arrives they have to share the affection of their partner with another individual. The response of a Neotype parent towards the arrival of a new child can be much like that of an older brother or sister – a combination of delight at having a new playmate mixed with resentment at the added competition and burden.

As with their relationships with their partners, the relationships that Neotypes have with religion can also be an uncomfortable combination of dependence and rebellion. Like rebellious youth, they are torn between a deep-seated need for guidance, through their insecurity, but at the same time they crave independence. It is a very fine balance and, depending on the individual, can swing either way. Neotypes are either deeply religious or fiercely rebellious as they explore alternative corporal ways of satisfying their infantile insecurity. Whatever the outcome, the Neotype is highly dependent on others, spiritual or real, for reassurance.

THE ULTRATYPE

Taken to its furthest extreme, the infantising process finally gives rise to the most extraordinarily imaginative yet insecure animal on this planet. The Ultratype represents the most distilled expression of those features that were central to the emergence of our species from our ancestors. But most revolutionary of all, in taking the infantising process to its extreme, evolution has actually created a totally novel feature, not only among humans but within the whole of the animal kingdom – true homosexuality. For the first time, there exist individuals who fail to develop beyond the immature stage of forming bonds exclusively with members of the same sex. Although same-sex bonding is a stage that all young primates – indeed all young humans go through – such is the extent of infantising in our species that a small percentage of people fail to develop beyond this point. Far from being adaptive, in terms of a distinct strategy to pass on more genes to the next generation, this behaviour is merely a by-product of our species' increasingly infantile condition.

The Ultratype is in many ways the most difficult type of human to live with. The emotional outbursts of Ultratypes are just one of many expressions of their insecurity. Even more than Neotypes, Ultratypes are very childlike. They are more volatile and can quickly change from being extremely affectionate to extremely hostile, often for the simplest of reasons. With maturity comes a self-assurance, a stability and a predictability that is lacking in the Ultratype. Frequently, Ultratypes take offence more easily over matters that they consider important to them, and require more emotional support than the other types of human. It is this extreme need for emotional support that has led to so many opting for a life devoted to religion. In God,

Ultratypes find the ultimate dominant father-figure to comfort their strong feelings of insecurity. Whereas the Neotype has something of a conflict between insecurity and independence, the Ultratype's high degree of infantile insecurity and dependence means that they are far more attracted to religion.

Ultratypes' relationships also face a unique problem. Whereas other types of human can choose partners who belong to any other type – with a Neotype man, for example, choosing a more mature Bureautype women, or vice versa – Ultratypes necessarily have to pair up with someone who is very similar to themselves. Because Ultratypes form pair-bond relationships only with others of the same type, they are precluded from pairing up with a more mature type of human who can provide them with the pseudo-parental care that their insecurity craves. The problem is clearly likely to be less severe in female Ultratypes, because of their resilient maternal qualities – although a recent survey found that only 30 per cent of lesbians wanted to have children – in male Ultratypes, the lack of nurturing feedback from their partner could certainly present problems.

Ultratypes' strong feelings of insecurity can not only lead to extreme levels of dependence, but also extreme levels of loyalty. Consequently, despite their lack of mature belligerence, they can make fearsome allies. In his *Symposium*, Plato recognised the loyalty of this type of person and the immense strength that this can produce:

> And if there were only some way of contriving that a state or army should be made up of [same-sex] lovers and their loves, they would be the very best governors of their own city... For what lover would choose rather to be seen by all mankind than by his beloved, either when abandoning his post or throwing

away his arms? He would be ready to die a thousand deaths rather than endure this. Or who would desert his beloved or fail him in the hour of danger? The veriest coward would become an inspired hero, equal to the bravest, at such a time; Love would inspire him.

The celebrated Theban Band, long supposed invincible, consisted of pairs of lovers fighting side by side, while as previously mentioned the Samurai of medieval Japan were renowned for going into battle alongside their lovers. There are also cases of fiercely loyal homosexual women, such as the 'Amazon' warriors of the Tupinamba Indians of north-eastern Brazil. Among the world's great military leaders, Alexander the Great is perhaps the most celebrated homosexual leader of all times. When his lover and general Hephastion died of fever, Alexander was so distraught that he designed a vast monument built from stones taken from the walls of Babylon to honour his beloved.

More so than with the other types of humans, there is a greater difference in the characteristics of male and female Ultratypes. This is because of the considerable differences between the social strategies of males and females during their formative years. With humans, as with many other primates, whereas young males establish extensive relationships with each other that will last for life, young females form small numbers of close relationships as they prepare to emigrate to join another group. Only in the later stages of social development, after they have transferred to their new group, do females develop the more extensive political and social relationships that are seen in young males. Since Ultratypes remain at the stage before emigration occurs in females, female Ultratypes are far less gregarious than male

Ultratypes. Unlike male Ultratypes, female Ultratypes are commonly highly cautious of members of the opposite sex, a behaviour that is associated with the incest-avoidance behaviour of young female primates who breed outside their birth-group. Not only do mature males pose a physical threat to young females, they are also likely to be very closely related. While the cost to a male of an incestuous mating would be minimal, the cost to a young female could be enormous. For both physical and sexual reasons, therefore, young female primates should be extremely wary of mature males.

Feeling comfortable with close physical contact is another quality of Ultratypes that results from their strong retention of infantile characters into adulthood. Consequently, Ultratypes are often attracted to professions that involve body contact, such as dealing with young children, and working in health-care. They are also extremely aware of small details of behaviour – this being a vital characteristic of the young to make up for their lack of physical strength, and which enables them to predict threatening situations before they arise – and as a result can be highly sympathetic. For this reason, Ultratypes are often more successful in professions that involve dealing with people at an intimate level than the more mature types of our species. Equally as noticeable, however, is the Ultratypes' creative talents and their attraction to the arts. An acute awareness of their surroundings is often reflected in their strong aesthetic sensitivity and success in professions that include a detailed appreciation of form and colour, including design and decoration.

Whether it is in the realm of acting, composing, painting, mathematics or science, some of our species' greatest achievements have been made by Ultratypes. Indeed around a hundred years ago, Henry Havelock Ellis noted:

In persons of genius of either sex there is a tendency for something of the man, the woman, and the child to coexist. It is not difficult to understand why this should be so, for genius carries us into a region where the strongly-differentiated signs of masculinity or femininity, having their end in procreation, are of little significance.

epilogue

It seems only right and proper at this late stage to expose myself to scrutiny and openly consider to which of the four categories I best belong. As it happens, I am in no doubt whatsoever about my level of infantising. I am a Neotype – my mother and father are Neotypes, and so am I. As with so many other Neotypes, I dream too much rather than getting things done. I have always found it very difficult to concentrate on anything for long enough to get it finished, and become distracted very easily. I relish novelty for novelty's sake – it does not have to be better, but simply different. Like Toad in *The Wind in the Willows*, I have a capricious alacrity that relentlessly drives me from one half-finished activity to another. Along with my children, I am emotionally dependent on my wife – a Bureautype – who copes admirably with the rest of the family's infantile mood swings, and brings us all to order when necessary. Like my children too, I want to do my own thing and not what is necessarily best for the others. Even more than them, I am rebellious. It does not matter what area of life it concerns, I feel compelled to question what others prescribe. As much as I crave the comfort that religion provides, I cannot bring myself to believe in a God of which I have no proof myself. I avoid conflict at nearly any cost. In true Neotype style, more than anything else, I would rather play – with words, with ideas, with bits of wood, or the family dog – until, that is, I get bored and want to move on to something else. As much as I adore both my son and daughter, I have never really felt myself to be old enough, or ready, to have children. It is an utterly ridiculous state of affairs for a forty-year-old mature mammal. But there again, I am not mature, and that is the whole point.

We have now reached the end of our journey. So, what more is there left to say? Perhaps merely to question whether our species' regression into a permanently infantile condition has been an unmitigated success. Of this, I am not too sure. Yes, it is true that our particular evolutionary route has inadvertently bestowed on us a mental capacity that is so superior to that of all other animals that we can now achieve global domination with unprecedented ease. Yes, it is true that our playful qualities have resulted in technological and artistic achievements of extraordinary variety and complexity. Without being touched by the infantising wand of evolution, there would have been no William Shakespeare, Albert Einstein, Wolfgang Amadeus Mozart or Charles Darwin. And yes, it is true that we have the potential to be exquisitely loving and tender in our relationships with our sexual partners.

But what of our seemingly infinite capacity to form large cooperative groups? How has this quality contributed to our species' well-being since it first evolved over six million years ago to protect us against ferocious carnivores? While it has undeniably allowed us to form highly complex and mutually supportive communities, in which people can pursue and develop their individual talents, this particular talent has a distinctly pernicious side to it. Through our ability to create massive armies that operate more like a single vast super-organism, we have been able to wreak havoc throughout the world and inflict horrendous misery on members of our own species like never before. Along with being the most playful species of all time, we are the most insecure and easily led. Rather than cherishing free will, humans seem to leap at the chance of surrendering it whenever they can. With our desperate need to identify a dominant 'parental' figure throughout our lives, from whom we expect protection in return for our childlike obedience, we all too willingly form religious crusades and national armies. In Faustian deals of the most depraved and nefarious kind, we destroy tens of thousands of fellow humans, merely to ingratiate ourselves with our spiritual or national leaders.

The answer is perhaps to encourage us to become even more infantile – to purge what little aggression is left after six million years of selecting big babies. From now on, it is very much a matter of nurture – treat people like children, provide them with all that they need so that they do not have to struggle to survive, and they will repay you by being childlike, creative and highly sociable. At last, we will possess infantile playfulness and tolerance without the lingering vestiges of our mature, aggressive, and antisocial past. Furthermore, lest xenophobia should sneak in alongside infantile insecurity, barriers between groups need to be broken down. The only way of expunging fear and distrust of others is by creating a perception of everybody belonging to one huge super-group.

There is, however, one major biological problem with such an extremely infantile scenario. Reproduction – or rather lack of it. As our species becomes increasingly infantile, we fail to develop the urge to have children. Already, some developed countries are in population freefall as increasingly more of their inhabitants fail to develop the mature urge to reproduce and instead remain playfully immature for much of their reproductive lives. The more people are encouraged by their governments to behave like dependent children, the more they want to have fun rather than babies. At its furthest extreme, if the whole population becomes sufficiently infantile, it will end up exclusively homosexual, in which case reproduction becomes a real problem. Yet again, our infantile nature is both our species' greatest strength and greatest threat. What a pickle. Oh well, time to take my little, hairless, bulbous-headed body out into the sun. Shall we play?

further reading

For those who might wish to read further on any of the subjects mentioned in this book, the following are some suggestions. Although the bookshops are replete with an astonishing array of highly erudite books on the subject of human biology, especially our evolution, these are some that have particularly captured my imagination.

Bee, H.L., *The Developing Child*, 9th ed (Allyn & Bacon, Boston, 2000)

Blakemore, C., *The Mind Machine* (BBC Books, London, 1988)

Blumenfeld, W.J. and Raymond, D., *Looking at Gay and Lesbian Life*, updated and expanded edition (Beacon, Boston, 1993)

Botting, K. and Botting, D., *Sex Appeal: the art and science of sexual attraction* (Boxtree, London, 1995)

Burr, C., *A Separate Creation: how biology makes us gay* (Bantam Press, London, 1996)

Buss, D.M., *The Evolution of Desire: strategies of human mating* (Basic Books, New York, 1994)

Calvin, W.H., *How Brains Think: the evolution of intelligence* (Phoenix, London, 1996)

Campbell, A., *The Opposite Sex* (Andromeda, Oxford, 1989)

de Waal, F. and Lanting, F., *Bonobo: the forgotten ape* (University of California Press, Berkeley, 1997)

de Waal, F.B., ed., *Tree of Origin: what primate behavior can tell us about human social evolution* (Harvard University Press, Cambridge, Mass., 2001)

Dennett, D.C., *Consciousness Explained* (Penguin Books, London, 1991)

Diamond, J., *The Rise and Fall of the Third Chimpanzee* (Vintage, London, 1991)

Diamond, J., *Why Sex is Fun: the evolution of human sexuality* (Phoenix, London, 1997)

Dixson, A.F., *Primate Sexuality: comparative studies of the prosimians, monkeys, apes and human beings* (Oxford University Press, Oxford, 1998)

Dunbar, R., *Grooming, Gossip and the Evolution of Language* (Faber and Faber, London, 1996)

Dunbar, R., *Primate Social Systems: studies in behavioural adaptation* (Croom Helm, London, 1988)

Eibl-Eibesfeldt, I., *Human Ethology* (Aldine de Gruyter, New York, 1989)

Fisher, H., *Anatomy of Love: the natural history of monogamy, adultery, and divorce* (Touchstone, New York, 1992)

Foley, R., *Another Unique Species: patterns in human evolutionary ecology* (Longman Scientific and Technical, Harlow, 1987)

Garber, M. *Vested Interests: cross-dressing and cultural anxiety* (Penguin Books, London, 1992)

Gazzaniga, M.S., *Nature's Mind: the biological roots of thinking, emotions, sexuality, language and intelligence* (Penguin Books, London, 1992)

Gibson, K.R. and Ingold, T., eds, *Tools, Language and Cognition in Human Evolution* (Cambridge University Press, Cambridge, 1993)

Gonsiorek, J.C. and Weinrich, J.D., eds, *Homosexuality: research implications for public policy* (Sage Publications, Newbury Park, 1991)

Goodall, J., *In the Shadow of Man* (Phoenix Giant, London, 1988)

Gould, S.J., *Ontogeny and Phylogeny* (Belknap, Harvard, 1977)

Gould, S.J., *The Mismeasure of Man* (Penguin Books, London, 1981)

Gribbin, J. and Cherfas, J., *The First Chimpanzee: in search of human origins* (Penguin Books, London, 2001)

Groves, C.P., *A Theory of Human and Primate Evolution* (Oxford University Press, Oxford, 1989)

Harris, J.R., *The Nurture Assumption* (Bloomsbury, London, 1998)

Hatfield, E. and Rapson, R.L., *Love and Sex: cross-cultural perspectives* (Allyn & Bacon, Boston, 1996)

Haviland, W.A., *Human Evolution and Prehistory*, 5th ed (Harcourt College, Orlando, 2000)

Jones, S., Martin, R. and Pilbeam, D., eds, *The Cambridge Encyclopaedia of Human Evolution* (Cambridge University Press, Cambridge, 1992)

Karlen, A., *Sexuality and Homosexuality: the complete account of male and female sexual behaviour and deviation – with case histories* (Macdonald, London, 1971)

Kingdon, J., *Self-Made Man and his Undoing* (Simon & Schuster, London, 1993)

Le Vay, S., *The Sexual Brain* (MIT Press, Cambridge, 1993)

Li, C.K., West, D.J. and Woodhouse, T.P., *Children's Sexual Encounters with Adults* (Duckworth, London, 1990)

Markale, J., *The Great Goddess* (Inner Traditions, Rochester, 1999)

Michael, R.T., Gagnon, J.H., Laumann, E.O. and Kolata, G., *Sex in America: a definitive survey* (Little, Brown & Company, Boston, 1994)

Miller, G., *The Mating Mind: how sexual choice shaped the evolution of human nature* (Heinemann, London, 2000)

Mithen, S., *The Prehistory of the Mind: a search for the origins of art, religion and science* (Phoenix, London, 1996)

Montagu, A., *Growing Young*, 2nd ed (Bergin and Garvey, New York, 1989)

Morgan, E., *The Descent of the Child: human evolution from a new perspective* (Souvenir Press, 1994)

Morris, D., *The Naked Ape* (Grafton, London, 1967)

Nitecki, M.H. and Nitecki, D.V., eds, *Origins of Anatomically Modern Humans* (Plenum Press, New York, 1994)

Parker, S.T. and Gibson, K.R., eds, *'Language' and Intelligence in Monkeys and Apes: comparative developmental perspectives* (Cambridge University Press, Cambridge, 1990)

Potts, M. and Short, R., *Ever Since Adam and Eve: the evolution of human sexuality* (Cambridge University Press, Cambridge, 1999)

Ridley, M., *The Red Queen: sex and the evolution of human nature* (Penguin Books, London, 1993)

Ridley, M., *The Origins of Virtue* (Viking, London, 1996)

Ruse, M., *Homosexuality: a philosophical inquiry* (Basil Blackwell, Oxford, 1988)

Serpell, J., ed., *The Domestic Dog: its evolution, behaviour, and inter-*

actions with people (Cambridge University Press, Cambridge, 1995)

Short, R.V. and Balaban, E., eds, *The Differences Between the Sexes* (Cambridge University Press, Cambridge, 1994)

Smuts, B.L., Cheney, D.L., Seyfarth, R.M., Wrangham, R.W. and Struhsaker, T.T., eds, *Primate Societies* (University of Chicago Press, Chicago, 1986)

Steele, J. and Shennan, S., eds, *The Archaeology of Human Ancestry: power, sex and tradition* (Routledge, London, 1996)

Thurer, S., *The Myths of Motherhood: how culture reinvents the good mother* (Houghton Mifflin Company, Boston, 1994)

Wilson, E.O., *On Human Nature* (Penguin, London, 1978)

Wrangham, R. and Peterson, D., *Demonic Males: apes and the origins of human violence* (Bloomsbury, London, 1996)

index